CROSSING BORDERS, STRETCHING BOUNDARIES

Crossing Borders, Stretching Boundaries

THE BOSE-EINSTEIN LECTURES ON INDO-GERMAN COOPERATION IN SCIENCE, TECHNOLOGY AND ENVIRONMENT

Edited by
STEPHAN F. VON WELCK

Consulting Editor
DAGMAR BERNSTORFF

MANOHAR
2000

First published 2000

ISBN 81-7304-349-3

Published by
Ajay Kumar Jain for
Manohar Publishers & Distributors
4753/23 Ansari Road, Daryaganj
New Delhi 110 002

Typeset by
Kohli Computers
15/28 Geeta Colony
Delhi 110 031

Printed at
Print Perfect
A-23, Mayapuri, Phase II
Delhi 110064

Contents

Illustrations

Einstein at his house in Caputh near Berlin in 1931.
Courtesy: Rabindra Bhavan, Shantiniketan

6. Albert Einstein and S.N. Bose shared a talent and love for music: Einstein played the violin since his childhood.
Courtesy: Birla Industrial and Technological Museum, Calcutta
7. S.N. Bose was a master on the esraj.
Courtesy: Birla Industrial and Technological Museum, Calcutta
8. Get together after the 8th Bose-Einstein Lecture on 'Science and Religion'. The speaker, Prof. M.M. Joshi, Minister of Human Resource Development, Prof. A.M. Khusro, Dagmar Bernstorff, the consulting editor.
Photo: S.K. Sharma

between pages 133–4

9. Jawaharlal Nehru and Albert Einstein shared a concern for non-violence and world peace. The Indian Prime Minister met Einstein at Princeton in 1949.
Courtesy: Nehru Memorial and Library, New Delhi
10. In the 1920s Berlin was a vibrant centre of discoveries in physics. Five Nobel laureates met here in 1931 (*from left*) Walther Nernst, Albert Einstein, Max Planck, Robert A. Millikan, Max von Laue.
Courtesy: Stiftung Preussischer Kulturbesitz, Berlin
11. Ongoing friendship: S.N. Bose lectured at the Fritz-Haber-Institut in Berlin in 1953 at the invitation of Prof. Max von Laue.
Courtesy: Archiv zur Geschichte der Max-Planck-Gesellschaft, Berlin-Dahlem
12. Prof. Reimar Lüst, the President of the Alexander-von-Humboldt-Stiftung, fascinated the audience with his 9th Bose-Einstein Lecture on Comets. From left: Ambassador H.-D. Dieckmann, Prof. A.M. Khusro, Prof. R. Lüst, Stephan F. von Welck, the editor.
Photo: S.K. Sharma

Foreword

Since their inception in early 1997, the Bose-Einstein Lectures on Science, Technology and Environment have become a major element of Indo-German cooperation in science and technology.

Satyendra Nath Bose and Albert Einstein, these two men of genius in the world of physics, are among the earliest and certainly the most eminent protagonists of this cooperation. What had been at their time a somewhat informal scientific exchange and collaboration without the support of public agencies, funding mechanisms or Internet assisted data exchange, has since developed into a full fledged programme for Indo-German cooperation in education, science, research and technology, based on two governmental agreements, funded by the governments of both countries and monitored by an Indo-German Committee on Science and Technology. It covers today a large number of scientific disciplines such as information technology, biotechnology and genetics, health and medical research, oceanography and geosciences, advanced materials and production technology as well as space research and technology. In the framework of this rather broad co-operation a solid network has been established, not only of personal contacts, if not friendships, between German and Indian scientists, but also of vivid interaction between universities, research centres, and other public and private R&D institutions in both countries.

The Bose-Einstein Lectures are complementing this interactive transboundary cooperation at an intellectual level, focusing on policy issues and the exchange of views on fundamental problems caused by the ever increasing importance of science, technology and the protection of the environment for one's day to day life and the future of generations to come. I do hope that the publication of these lectures will contribute to developing a better awareness of this impact not only within the scientific communities in Germany and India, but also in the minds of those who are responsible for shaping the future science, economic, cultural, social and foreign policy.

I am grateful to all those who have contributed to the publication of the Bose-Einstein Lectures: to the Science Counsellor of the German Embassy and Editor of the book, Dr. Stephan F. von Welck, who initiated and organized this series of lectures; to the many eminent speakers who shared their personal views and experiences with a vivid and interested audience; to Prof. A.M. Khusro, Chairman of the Finance Commission and President of the Federation of Indo-German Societies in India, as well as to Ambassador Khub Chand, past President of the Federation, and Dr. R. Natarajan, Director IIT Madras, who chaired the lecture sessions; to Mr. Anand Singh Bawa, Honorary Secretary of the Federation of Indo-German Societies in India, who organized most of the lecture sessions in close cooperation with the administration of the India International Centre; to Dr. Dagmar Gräfin Bernstorff, who contributed with outstanding personal commitment to the publication of the lectures in her capacity as Consulting Editor; to Mr. Ramesh Jain, Manohar Publishers & Distributors, for printing and publishing the book; and last but not least to the German Ministry of Education and Research, which has generously provided funds for this publication.

New Delhi
February 2000

DR. HEINRICH-DIETRICH DIECKMANN
German Ambassador to India

Preface

The Embassy of Germany and the Federation of Indo-German Societies in India, working collaboratively, had started a new chapter in their activities by organizing a series of lectures called the Bose-Einstein Lectures on Science, Technology and Environment. These lectures are named after the famous Indian physicist Satyendra Nath Bose and the celebrated German physicist Albert Einstein, who were aware of each other's work in physics and had cooperated in some of their scientific ventures.

These lectures, about a dozen in number, have attracted a different kind of audience than the normal run of lectures organized at times either jointly or separately by the German Embassy and the Federation of Indo-German Societies in India. The scientific community in Delhi, with some representation from other parts of India and even abroad, got deeply interested in these lectures, which while focusing on scientific themes, ranged from an explanation or elaboration of scientific theories and ideas in simplified terms to the impact of scientific and technological developments on social, economic and even political spheres. Eminent scientists from India and Germany undertook to deliver these lectures, and representatives of research institutions, companies engaged in R&D, and administrators as well as politicians attended these lectures with interest and enthusiasm. They are being presented in this publication and deal with a wide range of subjects from science to technology, from the past to the present and the future, and from evolution of scientific ideas to the futurology of environmental protection. The book is expected to receive wide attention. Apart from two distinguished Ambassadors of Germany to India, Ambassador Frank Elbe and Ambassador Dr. Heinrich-Dietrich Dieckmann, the

Federation of Indo-German Societies in India is indebted for these lectures as well as their publication in particular to Dr. Stephan F. von Welck of the German Embassy in India whose devotion to matters of science, technology and environment has been instrumental in bringing this work to see the light of day.

New Delhi
February 2000

PROF. A.M. KHUSRO
President, Federation of
Indo-German Societies in India

Introduction

Crossing borders, stretching boundaries: this seems to be the most remarkable feature of science and technology at the end of the twentieth century.

Science and technology have crossed national and cultural borders in an unprecedented way to become an essential part of today's emerging globalization. Bilateral cooperation in this field has developed from an incidental collaboration between individual scientists, based on their personal interests and initiatives, to a global network of well-defined and focused cooperation between government institutions, autonomous or corporate partners. Multilateral cooperation in science and technology, either within large international R&D programmes or through international organizations, has replaced national research activities in many fields, based on the necessity to share in the funding of the ever-increasing costs for large R&D projects and equipment.

At the same time science and technology have also crossed established borders between different scientific disciplines such as mathematics, physics or biology. Interdisciplinarity has thus become a hallmark of cutting-edge research programmes. Large research centres in which scientists and engineers from different scientific disciplines join their know-how, their experience and their equipment are nowadays the backbone of knowledge-driven national or international research programmes.

However, during the last four or five decades, science and technology have also stretched their traditional boundaries into areas which had so far been the exclusive domain of other human endeavours such as philosophy, ethics or religion. Unprecedented advances in knowledge and their application for example in nuclear physics, informatics or genetics, have confronted scientists and engineers with so far unknown fundamental questions regarding their activities. These questions touch the foundations of life, of human dignity and of the environment. The response to these questions will probably determine not only the future of science and technology, but also the future of human beings as a creation and of the planet on which they live.

The book *Crossing Borders, Stretching Boundaries* is based on lectures and articles by eminent scientists, engineers, administrators and politicians, which reflect recent developments in science and technology from the point of view of those who are personally involved in these developments. The subjects covered in the twelve lectures and two articles reach from the 'incidental' scientific cooperation between Albert Einstein and Satyendra Nath Bose, triggered by the personal initiative of the latter, to the globalization of R&D activities by multinationals like the DaimlerChrysler Corporation, the internationalization of space science and the impact of science and technology upon political power, ethics and religion.

Since the manuscripts and articles published in this book have been written with the objective to be read and understood not only by experts, but also by the interested public, I do hope that they will contribute to a learned interaction between scientists, managers, diplomats and policy makers on current and future developments in science, technology and the environment, which will most probably shape the life of generations to come more than most of us can imagine today.

New Delhi
February 2000

DR. STEPHAN FREIHERR VON WELCK
Science Counsellor, German Embassy

Satyendra Nath Bose and Albert Einstein: The Case of an Early Indo-German Scientific Cooperation

GISBERT F. ZU PUTLITZ

Science has a very high place in Indian and German culture. It has been pursued in both countries on a high international standard. Therefore scientists in both countries could learn from each other and cooperate for the benefit of both sides. Hence, one should not only take advantage of any opportunity to make this common potential visible to society and to politics, but also to the scientific community in both countries, to stimulate more intensive joint projects and endeavours.

Both, Albert Einstein and Satyendra Nath Bose were not only highly outstanding, internationally recognized and admired scientists but also personalities who should stimulate talented people. Both of them rose from modest beginnings to highest importance and fame. They truly serve as an example of encouragement to all young gifted people to pursue their education and scientific endeavour to the best of their abilities.

The doors Bose and Einstein have pushed open by sharing their knowledge remain relevant even today. 'The Bose-Einstein condensation' became a major scientific topic in the last two decades, but its foundation was laid in the twenties of the last century. It is, by the way, interesting to observe that at

*First Bose-Einstein Lecture, 12 February 1997, at India International Centre, New Delhi.

the beginning of the twentieth century the communication of scientific ideas and knowledge was not much slower than today and clearly very profound, while the Internet still had 60 years to become a reality.

Before I embark on Bose's and Einstein's scientific work I would like to draw a short life sketch of the two scientists. Satyendra Nath Bose was born on 1 January 1894 in Calcutta. He studied mixed mathematics at the Presidency College of Calcutta University from 1909 and received a Bachelor of Science degree in 1913, and a Master of Science degree in 1915 at the age of 21. In 1916 he became a lecturer at the University College of Science together with Meghnad Saha, an old friend of his since their college days who later became President of the Indian Institute of Science in Bangalore, an institution which has very close scientific ties with Germany today. In the same year C.V. Raman joined the University College of Science as Palit Professor of Physics.

In 1921 Bose was appointed as a Reader in Physics at the newly founded Dhaka University. His appointment nearly faltered when the university ran out of funds, but in the end Bose could stay and he wrote his famous paper 'Planck's Law and the Hypothesis of Light Quanta'. It was this paper which started Bose's interaction with Einstein who knew him from correspondence already because Bose and Saha translated Einstein's paper on General Relativity into English in 1916, when they were lecturers at the University College of Science in Calcutta. Incidentally this was the first translation of Einstein's work into English.

S.N. Bose initially sent his paper to the *Philosophical Magazine* in England and was very confident that it would appear there soon. However, this journal kept silent. Finally Bose sent the paper directly to Albert Einstein with a letter dated 4 June 1924 (see p. 26). He asked Einstein to accept this paper for publication in the *Zeitschrift für Physik* and did not hesitate to ask him to translate it into German. He wrote to Einstein 'because we are all your pupils, though profiting only by your teachings through your writings'. On 15 June Bose submitted his second paper: 'Thermal Equilibrium in the Radiation Field in the Presence of Matter' to Einstein.

Einstein understood immediately that Bose's paper meant a big progress for the theory of radiation. It was published in the August issue of the *Zeitschrift für Physik* under the title 'Planck's Gesetz und die Lichtquantenhypothese' with a note by Einstein at the end, saying: 'Bose's derivation of Planck's law signifies, in my opinion, an important step forward. The method used here gives also the quantum theory of an ideal gas, as I shall show elsewhere.' The next Bose paper was published a month later in the September issue of the *Zeitschrift für Physik*. In order to promote Bose's findings Einstein presented these papers in the Prussian Academy of Sciences on 10 July 1924 and on 8 January 1925. In both cases he extended Bose's method. Einstein's letter—actually a postcard—to Bose can be seen as a facsimile on page 27.

Dhaka University granted S.N. Bose leave of absence with full pay for two years, from 1924-6. He reached Paris in October 1924 and spent a year there working with Madame Marie Curie and with Maurice de Broglie.

One year later, on 8 October 1925, Bose arrived in Berlin and met Einstein in November 1925. Of course at that time the science atmosphere in Berlin was extremely inspiring. Bose wrote: 'I was very friendly with Franck, Einstein, Born, Ewald, Slizard and Mark'. At the end of his second year abroad (1926) Bose successfully applied for a professor's post at Dhaka University and was appointed Dean of the Faculty of Science at Dhaka University and full professor. Einstein had recommended him. Only after nearly twenty years and many offers to move elsewhere from Dhaka Bose returned to Calcutta as a Khaira Professor of Physics in 1945, a chair he held until 1956. He became President of the Indian Physical Society, President of the National Institute of Sciences, was made Fellow of the Royal Society and a national professor in 1959. He received many honorary degrees thereafter. On 4 February 1974 Bose died in Calcutta, one month after his 80th birthday had been celebrated throughout India.

Albert Einstein[1] was born on 14 March 1879 in Ulm, Germany, he was 15 years older than S.N. Bose. His academic career started in 1898 when he obtained a diploma in electrical engineering from the Eidgenössische Technische Hochschule (ETH) in Zürich and a school teacher diploma in mathematics and physics. From 1902 to 1909 Einstein worked at the Swiss Patent Office in Berne where he wrote in 1905 his famous papers on Planck's Law of Radiation, on Special Relativity, and on the mass energy equivalence. He received a PhD from Zürich University in 1905 and the Habilitation in 1908. Seven years later (1912) he published his paper on General Relativity. Following these outstanding achievements Einstein was called as Professor of Theoretical Physics to Prague University in 1910, but soon returned to his alma mater, the ETH, as Professor in 1912. He did not stay there for long as he was elected to the prestigious Royal Prussian Academy of Science in Berlin in 1913. Einstein received the Nobel prize in physics in 1921 for his work on the photo effect explained on the basis of quantum physics. It was only after this period, when Einstein was already famous for his groundbreaking discoveries, that Bose approached him in 1924. I think it speaks very much for the openness of Einste in that he took the papers by an unknown younger colleague from as far as Dhaka very seriously and immediately paved the way for Bose's success.

As one knows Albert Einstein was forced by the national socialists to leave Germany in 1933. Actually he choose not to return to Germany from a lecture tour abroad given the persecution of Jewish university teachers by the Nazis. He made a scientific home at the Institute for Advanced Study at Princeton University and stayed there until he died on 18 April 1955, at—the age of 76 years.

Einstein is a symbol for modern physics. His expulsion in 1933 is also a symbol for the intellectual downfall Germany

[1] See A. Fölsing, *Albert Einstein, A Biography*, New York, 1997; C.K. Majumdar et al. (eds), *S.N. Bose: The Man and his Work*, Part II, Calcutta, 1994.

experienced during Hitler's dictatorship. Einstein's support for the development of nuclear weapons must be seen against the background of his experience with the ruthless fascist despotism then ruling Germany. Scientists should never forget their responsibility to fight against the repetition of such a regime all over the world.

Clearly, my short review of the lives of Bose and Einstein is not comprehensive or complete in any respect. But one must understand, how important it was for Bose to interact with such an outstanding scientist like Einstein in Europe at the right time, when the development of quantum mechanics flowered with several important discoveries each year, and where the total atmosphere was so intriguing and stimulating for S.N. Bose.

But now I turn to the physics Bose and Einstein pursued in the papers mentioned already in Bose's curriculum, the scientific case for a different statistic in physics. This originates from the question, what energy states particles, for example gas molecules, can occupy at different temperatures.

The first approach to this problem came from Maxwell and Boltzmann. For two distinguishable particles x and y there would be four possibilities in the arrangement of them in two different energy states. Both particles could be in the lower state, or in the upper state, the x could be in the lower state and the y in the upper state, or vice versa. Hence, for two distinguishable particles there exist four possibilities. Without proof I mention that for three particles 27 possibilities do exist. This Maxwell-Boltzmann theory describes very well the energy distribution of particles at a given temperature. In connection with *photons* which are indistinguishable, Bose pointed out that only three different possibilities exist. Both photons can occupy the lower state or the upper state or one photon each populating the upper and the lower state. Consequently for two photons there are three possibilities, for three photons there exist ten. Einstein extended these considerations applied by Bose to photons to other particles as long as they have zero or even spin. In this case it would not be excluded by the Pauli principle that these particles could occupy the same quantum state with all their quantum numbers being equal. However, with Pauli's principle that one quantum state could be occupied only by two particles if they differ in one of their quantum numbers there are one or two possibilities for two particles respectively depending whether they are distinguishable or undistinguishable. This led to the so-called Fermi-Dirac statistic. The particles like electrons with their odd spin of S = 1/2 were named by Dirac *fermions*. Equally the particles with zero or even spin obeyed the Bose-Einstein statistics and were called *bosons* after Bose. The picture was now complete. The sum of the bosons and of the fermions constitute the universe!

The facts explained in the previous paragraph resulted in a completely different distribution of particles as a function of their energy or equivalent, the temperature. The Fermi-Dirac statistic at zero absolute temperature is very simple to understand: Each state can only accommodate two particles, the others are stacked from one state to the other until the particles are all accommodated. So the occupation number is two and remains so until it breaks down to zero at the highest energy. In the Bose-Einstein statistic the occupation number follows an exponential law for higher temperatures. In order to clarify the consequences of the new statistic one's attention is drawn to the two cases depicted for the different statistics. In the Bose-Einstein statistic most of the particles are in the ground state while the others are distributed to higher states ex-ponentially. Only if the temperature approaches zero all particles will be in the lower state. In the Fermi-Dirac statistic if the temperature is equal to zero all particles are in one particular state (actually two of them in each state) with increasing energy until all the particles occupy a state. With rising temperature this occupation may be slightly softened by the Maxwell-Boltzmann method at the edge of the distribution. However, basically the occupation pattern remains the same.

Bose particles, the bosons, occupy the lowest possible state at a temperature of T = 0°K, the zero point of the absolute temperature scale. Already Bose and Einstein suggested, that under these conditions condensation into a single macroscopic state could take place, a 'Bose-Einstein condensate', as it was called soon after their proposal. Such a condensate, a macroscopic quantum mechanical state, requires not only that all particles are in the lowest single level but also linked together to a common phase, like a freight train, where all the identical lorries have the same direction and velocity, i.e. an identical momentum.

Now one may ask whether the case for the temperature being zero degree Kelvin where all atoms are in the lower state could lead to a condensate, in other words to a macroscopic condensate where all the particles move together. In order to obtain such a condensate an additional condition has to be fulfilled. This condition is called phase coherence. A picture with five pendulums would show what I mean. All the pendular move at the same time in the same direction keeping their distance constant. This long phase coherence is equal to a 'broken gauge symmetry', the matrix for the occupation number and phases between the particles have diagonal and off diagonal elements, the first ones being the occupation numbers while the latter ones the phase between them. These few remarks of course are meant to address the physicists among the readers.

London proposed more than 50 years ago that the presence of super-fluidity in liquid He at temperatures below Tc < 2.17 K at normal pressure of 1 bar is linked with a possible 'Bose-Einstein condensation'. Superfluidity is a phenomenon where a liquid looses friction completely, it can creep through the smallest holes or even flow over the edge of a beaker. Today it is known that only 9 per cent of the liquid He at T = 0 K are in a Bose-Einstein condensate.

It was only in the last ten years that a new chapter in the exciting history of the 'Bose-Einstein condensation' was written. This had to do with the development of atom traps which were suited to hold larger numbers of atoms in the micro Kelvin (μ°K) to nano Kelvin (n°K) range (10–6 to 10–9°K) in a tiny volume. There 'Bose-Einstein condensation' was observed for up to 100 per cent of the interacting particles, for example Cs, Rb or Na atoms. These discoveries are important for the understanding of the interacting bosons, but they also have a tremendous impact on possible applications like time standards, ultra precise measurements of natural constants and their variation with time and location, etc. The expulsion of atom by atom from a Bose-Einstein condensate in a trap realizes the picture of the forementioned atomic freight train, so to speak, i.e. an atom laser.

Summarizing again it could be said that bound systems with zero or even spin are subject to possible 'Bose-Einstein condensation'. This principle governs the most modern physics. Pairs of electrons with spin up and down could form cooper pairs which are considered to be responsible for superconductivity. Equal considerations are true for the bound state of an electron and a proton with opposite spin directions, i.e. the lowest hyperfine structure state in hydrogen or in other atoms where the nuclear and the electron spin can form an integer number. Bosons can also be formed between neutrons and protons and of course between quarks and anti-quarks as well as in pion and kaon condensates. Last not least the existence of the x-boson is proposed as a condensate of the top quark and the anti-top quark. These remarks should only be a demonstration that bosons and Bose-Einstein condensates are key elements in the most modern theories in physics.

Bose and Einstein are an example of an Indo-German scientific cooperation of fundamental impact. Why did this joint effort take place in such a successful way? Because all the necessary ingredients were there at the right moment! The right passion for the understanding of nature, a common language (English and Mathematics), the freedom to pursue this research leading to progress in the future not influenced by narrow-minded funding agencies aiming at ephemeral technology transfer.

It is 80 years since Bose first submitted his papers to Einstein. And only now applications emerge from Bose-Einstein condensates which may lead to an unprecedented precision in metrology, that enables one to ask questions and answer them, e.g. on the origin of the universe, its development as a function of time, on the forces holding it together, and on the basic constituents of matter. In this way, the amalgamation of the ideas of Bose and Einstein can be understood as a crucial tool for the noblest obligation of all scientists: To search for the ultimate truth.

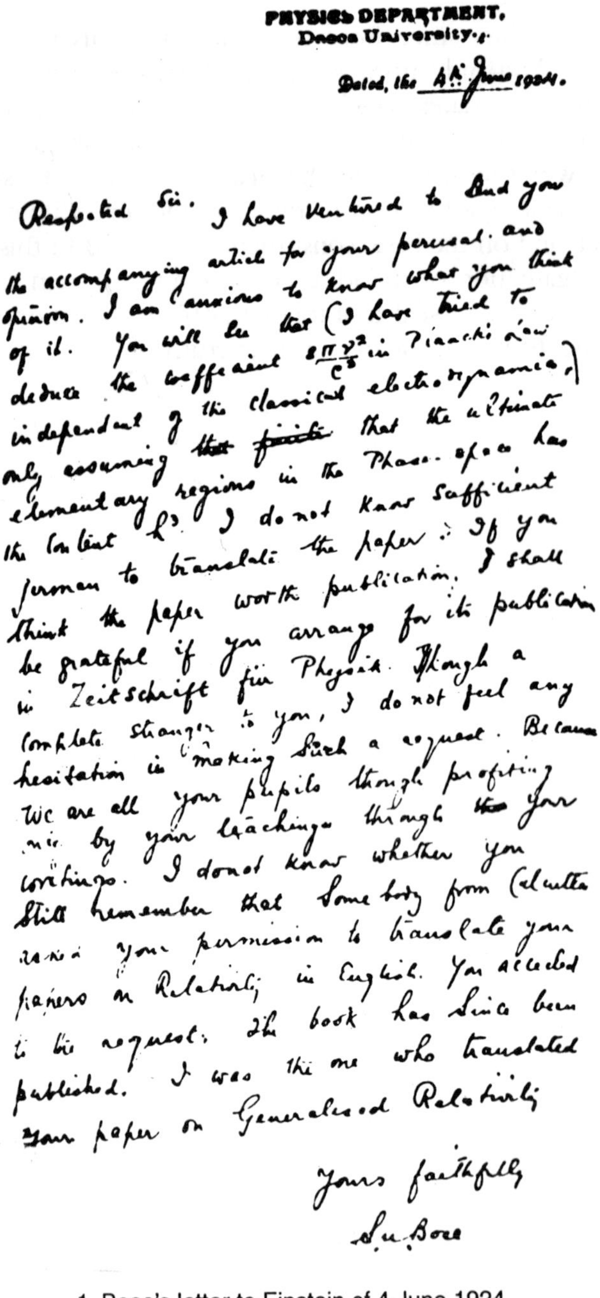

PHYSICS DEPARTMENT,
Dacca University.

Dated, the 4th June 1924.

Respected Sir, I have ventured to send you the accompanying article for your perusal and opinion. I am anxious to know what you think of it. You will see that I have tried to deduce the coefficient $\frac{8\pi\nu^2}{c^3}$ in Planck's Law independent of the classical electrodynamics, only assuming that the ultimate elementary regions in the Phase-space has the content h^3. I do not know sufficient German to translate the paper. If you think the paper worth publication, I shall be grateful if you arrange for its publication in Zeitschrift für Physik. Though a complete stranger to you, I do not feel any hesitation in making such a request. Because we are all your pupils though profiting only by your teachings through your writings. I do not know whether you still remember that somebody from Calcutta asked your permission to translate your papers on Relativity in English. You acceded to the request. The book has since been published. I was the one who translated your paper on Generalised Relativity.

Yours faithfully
S. N. Bose

1. Bose's letter to Einstein of 4 June 1924.

2. VII. 24

Lieber. Herr Kollege!

Ich habe ihre Arbeit über-
-setzt und der Zeitschrift für
Physik zum Druck übergeben.
Sie bedeutet einen wichtigen
Fortschritt und hat mir
sehr gut gefallen. Ihre
Einwände gegen meine Arbeit
finde ich zwar nicht richtig.
Denn das Wien'sche Ver-
schiebungsgesetz setzt die
undulations theorie nicht voraus
und das Bohr-che korrespondenz-
prinzip ist uberhaupt nicht
verwendet. Doch dies thuts
nichts. Sie haben als erster
den Faktor quantentheoretisch
abgeleitet wenn auch wegen
des Polarisations-Faktors 2
nicht ganz streng. Es ist
ein schöner Fortschritt.

Mit freundlichen Gruss
Ihr A Einstein.
(Sd)

2. Einstein's postcard to Bose of 2 July 1924.

Einstein

SATYENDRA NATH BOSE

All must have surely heard the name of the great genius Einstein. His biography has been written in various languages and has also been translated into Bengali. If one is interested, one should look for them and read them. Einstein was born on 14 March 1879 in a Jewish family in the city of Ulm in southern Germany. His father, Hermann, started his business with a small capital. After trying his luck in various places, he set up a small shop of electrical goods[1] first in Munich and then in Milan, Italy. The income from it was meager and gradually his savings were all gone to pay off the debts. Therefore, when son Albert finished his school in Munich the family found it beyond their means to bear the expenses of his higher studies in a German university. Einstein then applied for admission to the Polytechnikum, the famous technical college in Zürich in Switzerland. His application was rejected the first time. He got excellent marks in mathematics in the entrance examination. But because of his inadequate knowledge of biology and foreign languages, he had to study these subjects in a Swiss school for one more year. Finally he was admitted in 1896 and studied in the Polytechnikum[2] for four

*First published in Bengali in *Adi Mahakali Pathshala Patrika*, March 1965, the English translation published by S.N. Bose National Centre for Basic Sciences, in S.N. Bose, *The Man and His Work*, Vol. II, Calcutta, 1994, pp. 285–9.

[1] Father Einstein and his relatives founded the Electrotechnische Fabrik J. Einstein & Cie in Munich which produced bulbs, lamps, telephones and provided electrification. When they did not get the contract to electrify parts of Munich the firm ran into difficulties (editor's note).

[2] Later the Eidgenössische Technische Hochschule, ETH (editor's note).

years. He read what he liked in physics, like the original works of the great Maxwell, Boltzmann, Hertz and Kirchhoff and mastered them. He completed his studies at the Polytechnikum in the year 1900. Initially, he had to do various odd jobs, like giving private tuitions, to earn a living. Eventually he got a job in the Swiss Patent Office in Berne in July 1902 on an annual salary of 3,500 Swiss francs.

Writing about this small job, Einstein remarked towards the end of his life:

> This job at the beginning of my career seems to me now to have been a blessing in disguise. It proved to be of better value to me than teaching in a school or college, because I had a lot of free time to think independently about different areas of science.
>
> Young scientists of my age who got the opportunity to do scientific research in schools and colleges right after getting degrees had to remain occupied with inconsequential problems in order to produce quick results. The firmness of character and will power that are necessary to resist the temptation of easy success and to dedicate oneself wholeheartedly to the search of the fundamental and profound theories of science, are indeed rare among young scientists. So they get lost in small things.

Einstein devoted all his free time to the pursuit of science. The goddess of learning was pleased by his silent and devoted work for three long years. He published his first paper on the relativity theory in 1905 at the age of twenty-six. A certain problem in the theory of light was then baffling scientists all over the world. His original paper threw new light on the subject from a new standpoint and revolutionized the world of science. Veteran scientists like Poincaré, Lorentz and Planck had the highest praise to offer to this budding young scientist. He could not live in the small city of Berne for long. The offer of a professorship first came from Prague. There his original papers on various subjects elevated him to the rank of world-famous scientists, and he received an invitation to teach in Berlin just before the First World War.

Einstein had many new things to say. He showed how it was possible to give a causal description of the events of the external world independent of the state of motion and position of the observer. He also said many novel things about the new standpoint that a scientist requires in order to investigate the world on the assumption that a description of the external world independent of the observer is possible. One has learnt from Einstein many new facts about matter, energy, force and gravitation, and following the path charted out by him, several scientists of the new era have progressed further and earned fame. The new hypothesis that Planck proposed in 1900 about the interaction of light and matter and about exchange of action-energy, has gone through various modifications and refinements to evolve into the modern quantum theory. In the papers that he wrote in support of his theory, Einstein added a new element to Planck's idea—he introduced the idea of light-quanta or photons. With its help several mysteries were clearly and easily solved. Although it became quite impossible to reject the wave theory and to think of light, as Newton did, as a collection of light corpuscles, Einstein thought deeply about how these completely contradictory properties could be present in light.

These new theories surfaced when one was still in college, but the First World War began before these theories got well circulated in this country. One learnt about these fascinating theories much later.

After discovering the relativity theory, he spent ten years thinking about gravity and how it could be placed in the context of his relativistic space-time theory. Finally, he explained gravity in a totally new way. However, scientists in different countries could not pay much attention to it because of the War. Einstein had predicted that according to his theory light rays would be slightly deflected when they graze past the sun's surface. He even calculated the extent of this deflection. Germany was defeated in 1918, and peace was restored in the world. Then scientists the world over found time to test Einstein's theory of gravitation.

In 1919 a sensational news astounding all scientists appeared in newspapers all over the world. During the full solar eclipse that year a group of British scientists tried to find out whether light from a distant star got slightly deflected by the gravitational pull of the sun while passing close to it. The extent of the deflection was expected to be 1.75 second of arc according to Einstein's new theory. The scientists were amazed to find that the results of their observation almost fully supported Einstein's calculation. Einstein's fame spread all over the world from this event.

One had only just begun to teach at that time. The Calcutta University had started postgraduate classes in science. Three of us, Dr. Meghnad Saha, Professor Prasanta Mahalanobis and I selected the main papers on the new theories and translated them from the original German into English, and the Calcutta University published them as a book. Initially this book was the only one in which English translations of the main papers on relativity were available. Other translations followed and this book is now no longer available.

The Dhaka University was established in 1921. I had to go there as a Reader in the Physics Department. I had to explain many ideas to the higher classes. During that time I wrote a paper synthesizing the basic principles of Planck's theory and Einstein's light-quantum (photon) hypothesis. I wrote it in English and sent it to England for publication. Meanwhile, I was also curious to know Einstein's opinion of the new idea. So I also sent a copy of the unpublished paper to him. I did not then hope for any unexpected result—in fact, I had very little hope that I would be able to draw his attention to this matter at all.

Yet an amazing thing happened. Einstein sent his reply on a postcard (see pp. 26, 27) within a few weeks of receiving my paper. He acknowledged that this helped to take his theory further. He added that my views were so valuable that he would himself translate the paper into German and get it published. His translation was published (in 1924), long before anything was heard from London.

The possibility of my going abroad was then being discussed.

Professor Einstein's letter put an end to the controversy within the Executive Committee of Dhaka University regarding sending me abroad on deputation.

I got both leave and travel expenses from the University to go abroad for two years. I got my visa for Germany quite easily, thanks to the little postcard from Einstein. This letter opened up a new phase in my scientific career. Later when I arrived in Europe, I found that practically everybody had read and was discussing my paper. Einstein himself had applied my statistical method to material particles and extended the sphere of its application considerably. As long as I was in Germany, I had easy access to all the scientific institutions. In addition, like German professors I enjoyed the privilege of borrowing books from the government libraries and read them at home without depositing any money— all this was because of the guru's blessings.

I arrived in Berlin in 1925. The Einsteins then lived in a small flat on 5 Haberland Strasse. I was fortunate to be able to visit him often and have long discussions with him on many matters. I will conclude by relating an interesting conversation I had with him one day. After the World War the Jews got an opportunity to establish a new State in the land of their ancestors. A new university was set up in Jerusalem at that time. Einstein went to deliver an address at this new and obscure university. He also went to Japan more or less at the same time. He had an invitation to visit India but for some reason could not come here. The Jewish State of Israel was then growing under British protection. Many thought that this link with the British would be entirely for the good of the Jews. Einstein also held the same view. On the other hand, many did not fully support the British policy. A lot of Muslim Arabs used to live in Palestine then. They did not welcome the arrival of the Jews. The British rulers sought to establish their domination on a divide and rule policy just as they did in this country. Everybody in India is quite familiar with this policy, and one is still suffering from its evil effects.

The time I am talking about was when the movement to

drive the British out had already started in this country. The revolutionaries were making futile attempts to drive the British away by armed struggle, and the Congress was raising the slogan of non-cooperation. All these appeared rather strange to Einstein.

One day he confided to me, 'I think Englishmen are better than other Western colonial nations, and I feel that they are far better than the French and the Dutch. You should not be surprised at a German like me praising the English (after the World War). Now, tell me, do you really want that the British should quit your country?' I said, 'Of course, we all want to determine our destinies ourselves'. He was not quite convinced. He raised a hypothetical question. He said, 'Suppose, there was a button near you and all the Englishmen would quit India if you were to press it. So, would you press the button?' I smiled and said, 'If God were to grant me such an opportunity, I will not hesitate even for a moment to press it'. 'Really?', he said, and kept quiet for a while. I asked him, 'Well, why do you Jews then want to establish a new Israeli state? Even you seem to be fairly inclined in its favour'. He said, 'I have, of course, now understood what you are saying—it is an emotional matter and cannot be explained by logic'. Then, on my way back to India I met another world-famous man in France. He had visited this country. He said, 'Why create yet another kind of nationalism led by all these emotions? It could only aggravate conflicts and unrest. The Jews will separate as a different race and establish an independent nation—it is difficult to say what the result will be.' It was clear that opinions differed among great men. Of course, other French Jews did not agree with this French pundit. That was forty years ago.

Afterwards, both India and Israel became independent states almost at the same time. Unfortunately, the Englishman's divisive policy had sown the seeds of fatal racial conflict in the wombs of these two young states. Nobody can say what the final outcome will be.

I left Germany in the middle of 1926. Then anti-semitism

took a serious turn in Germany. Einstein had to leave his own land and take shelter in Princeton in America before the start of the Second World War.[3] It was from there that he propagated his latest theory. He stood for world peace, and he prayed for it in unambiguous terms. His immortal words about Gandhi will remain indelible in Indian history.

I was not destined to meet Einstein again. One corresponded from time to time. I was present in Berne in 1955 on the occasion of the golden jubilee of the relativity theory. But a few days before that, Einstein had suddenly expired in a hospital at Princeton. The Berne celebration had to take place without him.

[3]In 1932 Einstein did not return to Germany from a lecture tour abroad (editor's note).

Plate 1: S.N. Bose in his study, Einstein looks on. The portrait is still hanging in the family's home in Calcutta.
Courtesy: Birla Industrial and Technological Museum, Calcutta

Plate 2: Albert Einstein in his study, the 'tower', in Berlin where he thought, wrote and talked to S.N. Bose (1928).
Courtesy: Stiftung Preussischer Kulturbesitz, Berlin

Plate 3: Young Professor S.N. Bose at Dhaka University in 1928, after his return from Europe.
Courtesy: Birla Industrial and Technological Muscuin, Calcutta

Plate 4: Dr. rer. nat. Gisbert F. zu Putlitz, Professor of Physics at Heidelberg University, delivered the 1st Bose-Einstein Lecture on 12 February 1997. From left: Ambassador Frank Elbe, Prof. A.M. Khusro, Prof. Gisbert F. zu Putlitz.
Photo: S.K.Sharma

Bose-Einstein Condensation: From Theory to Experimental Observation to Application

GERHARD REMPE

All over the world, research is being directed towards the study of extremely cold gases. One of the most spectacular phenomena is the Bose-Einstein condensation, in which all the gas atoms form a common matter wave. Some 20 research groups are now able to produce such condensates worldwide, including research scientists in Constance, now at the Max-Planck-Institut für Quantenoptik at Garching.

After decades of effort, scientists have recently succeeded in observing a phenomenon in a gas of ultra cold atoms, which is attributable to the indistinguishability of identical particles. This indistinguishability contradicts our everyday experience and can be illustrated by the following example: If milk is poured into a cup of coffee it will soon mix completely with the coffee. The reverse process, separation, does not take place. Evidently an ordered state, in which the milk is separated from the coffee, will not occur spontaneously. But what happens when milk is poured into a cup which already has milk in it?

For a long time, nobody was able to say whether the disorder increases when identical substances are mixed together. On the one hand, everyone knows that no macroscopic effect is observable when two identical substances are mixed. On the other, by tracking the individual molecules it should be possible, at least in principle, to identify precisely from which cup a given molecule has originated. Hence, even for identical substances, the same situation should occur as in the example with the milk in the coffee.

*Courtesy: forschung. Mitteilungen der DFG 1–2/98.

Only the advent of quantum physics has brought an explanation of this mixing paradox, for in this case, one has to dispense with the concept of particle paths. Hence, it is impossible to identify the place of origin of the individual molecule once mixed. As a consequence, the molecules lose their identity. Thus in quantum mechanics, identical particles are in principle indistinguishable.

Light is an ideal gas

What consequence has this quantum aspect of indistinguishability? One answer to this question appeared from an unexpected direction. Albert Einstein had discovered that light consists of quantum particles, or photons. This idea was taken up by Satyendra Nath Bose, who realized that light was a photon gas. It is even an ideal gas as its rays of light could cross unhindered, with the photons not suffering any mutual interaction. Bose now assumed that photons are mutually indistinguishable, and are thus subject to different statistical laws than distinguishable objects. On the basis of this assumption he was able to describe an important quantum phenomenon: The spectrum of colours of light emitted from a glowing object.

At first, Bose's manuscript was not accepted for publication. So he sent it to Einstein with a request for his comments (see pp. 26, 27). Einstein recognized the novel value of the study and recommended publication. He also recognized the formal relationship between the photon gas and the ideal molecular gas. Einstein published two works on the 'Quantum theory of the ideal monatomic, ideal gas', in which he also described the phenomenon which was later to be called 'Bose-Einstein condensation'. He asserted that under certain circumstances 'something similar occurs to what happens when steam is compressed isothermally above its saturation volume. A separation takes place; one part condenses, the rest remains as a saturated ideal gas.'

Why is this phenomenon so fascinating? Is this effect not similar to what occurs when steam is condensed to water? Einstein replies that 'there is a maximum density of agitated molecules for each temperature. If this density is exceeded the excess molecules separate out motionless (condense without any attractive force),' and reminds one that condensation takes place in an ideal gas. Steam is not an ideal gas as there is mutual attraction between molecules which enables drops to form. In contrast to this, however, no forces act between the particles of an ideal gas. Hence, a Bose-Einstein condensate is also not liquid but gaseous. The condensed atoms are distinguished from the remaining atoms solely by the fact that they are no longer in motion! The sole cause of this macroscopic quantum phenomenon is the indistinguishability of the particles.

Under what conditions does condensation take place in an ideal gas? Quantum mechanics supplies a simple answer to this question: The collapse of the concept of particle path is closely linked to the fact that atoms behave like waves which, like water waves, propagate omni-directionally. However, at room temperature, the wavelength is much smaller than the atoms themselves, so that the wave nature is not evident. If a longer wavelength is required the gas must be cooled. At very low temperatures the situation can occur where the wavelength of the atoms is greater than the distance between particles themselves—a necessary precondition for Bose-Einstein condensation: When the waves of the adjacent atoms overlap, a common wave of matter can occur which extends over the entire volume of the gas.

However, in nature, two classes of particles, fermions and bosons, exist, which obey different rules for the population of quantum states: Fermions follow the Pauli principle, which will not permit a quantum state to contain more than one particle, whereas bosons, in contrast, permit more than one particle in a quantum state. Obviously, the Bose nature is also a precondition for Bose-Einstein condensation, in which a macroscopic portion of the particles adopt their minimum energy state. Electrons are fermions, but most of the atoms in the periodic table of the elements are bosons.

Cooling atoms with light

In the normal case, all gases become either liquid or solid at low temperatures. Cooling a gas to a stage where Bose-Einstein condensation occurs presents problems due to the fact that the simultaneous collision of several gas particles can cause nucleation. If this is to be avoided, it's density must be kept low. In terms of Bose-Einstein condensation, this means that very long wavelengths, in other words, very low temperatures have to be achieved—something which is not possible using conventional methods of low temperature physics. The breakthrough only came with the development of a completely new technology: Cooling atoms with light. The pioneering progress made in this field was acknowledged by the award of the Nobel prize for Physics in 1997.[1]

Bose-Einstein condensation observed

The basic idea of light cooling is simple: Cooling means that heat is withdrawn from the gas. To achieve this, the fact exploited is that atoms always emit light of a characteristic energy. If atoms are illuminated with light of a somewhat lower energy, they can certainly absorb the light, but on an average more energy will be emitted than is absorbed. The missing energy comes from the kinetic energy of the particles. And the slower the motion of the gas particles, the lower is the temperature. Using this method, absolute zero, minus 273 degree Celsius, can be approached to within a few millionths of a degree. The surprising fact is that the gas is not cooled by contact with the walls of the vessel, but only through the action of light!

[1]To William Philipps (USA), Steven Chu (USA) and Claude Cohen-Tannoudji (France) (editor's note).

Light cooling alone, however, is not sufficient to produce a Bose-Einstein condensate. This has so far only been achieved in magnetically stored gases whose temperature has been further lowered by evaporation cooling, where the atoms containing the highest energy levels are removed selectively from the gas. The remainder of the gas cools down—in a manner similar to that by which a hot soup is cooled by blowing on it. It was using this process that, in the course of a series of exciting experiments, three American teams at the Joint Institute for Laboratory Astrophysics in Boulder, at Rice University in Houston, and at the Massachusetts Institute of Technology in Cambridge, finally succeeded in observing Bose-Einstein condensation.

Since then, scientists all over the world have been furiously seeking to produce this novel material state in other laboratories, too. After only two years of preparatory effort, scientists at Constance managed to catch up with international developments in September 1997. In their very first experiment they succeeded in cooling a gas of rubidium atoms to a temperature of a hundred billionths of a degree Celsius, where over a hundred thousand atoms condense in a single matter wave no wider than the diameter of a human hair!

How does one recognize a condensate? In contrast to the atoms of a normal gas, condensed atoms are motionless. This suggests that the velocity of the atoms can be measured. To achieve this purpose, the gas is released suddenly by switching off the magnetic trap. Below one millionth of a degree two clouds can be observed expanding at different rates—one, a large isotropic cloud, whose extent is temperature dependent and which thus contains normal atoms, and the other, a smaller, central cloud of condensed atoms which only expands slowly and independently of the temperature. The gradual expansion of the condensate shows that the atoms are not actually at rest. In fact, the Heisenberg uncertainty principle for atoms localized in the magnetic trap leads one to expect a small velocity other than zero. As

Einstein correctly postulated, atoms can only come completely to rest in an infinitely extended gas. In further distinction to Einstein's theory, rubidium atoms only approximate to an ideal gas. In reality, the expansion of the condensate in the experiments was determined not by the Heisenberg uncertainty principle, but by the mutual interaction between the atoms.

The goal is a 'glowing' atom laser

One aim of future research will be to build a coherent source of matter waves. An atomic laser of this sort differs from conventional sources of atoms as a light laser does from an electric light bulb, in that it emits a monochromatic beam of atoms. At the Massachusetts Institute of Technology this has been achieved by briefly extracting the atoms from a Bose-Einstein condensate. The challenge is now to build a continuously 'glowing' atom laser.

Science and Religion

MURLI MANOHAR JOSHI

S.N. Bose and Albert Einstein both were great scientists, who not only made significant contributions but also gave a new dimension to scientific thought. Satyendra Nath Bose as a young scientist had sent a paper on photon statistics to the celebrated scientific journal, *Nature*, for publication. It's editor, in his wisdom, rejected the paper and pointed out that Bose omitted an important characteristic of photons. Bose then sent this paper to Einstein for his comments and in this omission by Bose, Einstein found a solution to a longstanding problem in the theory of light. Therefore, this paper is known as Bose-Einstein Statistics. It was the greatness of Einstein that he had only appended his name as the discussant even though Bose was a very young lecturer in India.

Prof. Bose's name stands on a very high pedestal in the field of theoretical physics. His work on Bose-Einstein Statistics, his prediction of bosons—intra-atomic particles not obeying Pauli's principle and condensation of matter at temperature approximating to absolute zero are the shining examples of his scientific genius. Now in 1995, S.N. Bose's prediction regarding the possibility of a *sui-generis* state of matter at a temperature infinitesimally small and close to absolute zero has been proved to be correct by the experiments performed at the University of Colorado.

Satyen Bose was not only a physicist, but he was also a versatile genius. He used to visit the University of Allahabad because one of my teachers, Prof. Kedareshwar Banerjee,

*Eighth Bose-Einstein Lecture, 27 October 1998, India International Centre, New Delhi.

was his student and I had the good fortune of remaining in attendance to this great genius. He used to come and stay in that very house at Allahabad in which I now live. So, I have a very special affiliation with S.N. Bose.

Prof. Bose could handle the problems of numismatics as easily as he could handle the problems of physics. He could discuss the problems of palaeobotany with as much ease and facility as he used to teach physics. He was well versed in ancient Indian literature and in Indian philosophy. He was a versatile genius and I offer my salutations to him.

At the beginning of the twentieth century, which also heralded a new era in physics, Einstein's intellectual contribution resulted in giving a new direction to science. His two articles published in 1905 initiated two revolutionary trends of thought. His 'Special Theory of Relativity' and an entirely new method of interpreting the electromagnetic theory were responsible for laying the foundation of modern physics particularly the Quantum Theory. His papers published in the early years of the twentieth century produced new concepts—a break from classical physics.

Einstein tried to integrate mechanics and electrodynamics. This resulted in the development of the 'Special Theory of Relativity' and produced revolutionary changes in the concept of space and time of Newtonian physics. According to Einstein space and time were not different entities. Space was not three-dimensional and space and time together constituted a four-dimensional space-time continuum. Thus one could not talk of absolute space and absolute time. Both space and time become symbols which the scientist uses to describe his experiences and observations. This leads to the inescapable conclusion that the entire framework of classical physics undergoes a radical change where not only space and time lose their meaning but also mass becomes a form of energy. Another fundamental postulate of Einstein's theory was the constancy of the velocity of light 'c' which occurred in the mass-energy relationship as $E = mc^2$. Einstein in 1915 attempted to include gravity in the ambit of his 'Special Theory of Relativity'. In this new framework gravity has the effect of curving space and time. This new theory known as 'General Theory of Relativity' propounds that wherever a body like stars possessing huge mass exists, the space around it becomes curved. Remembering that time is inextricably connected with space, it is as well affected by matter. In ultimate analysis space and time do depend upon the distribution of matter and the concept of empty space becomes redundant. Although the 'General Theory of Relativity' has not yet been fully confirmed yet, it remains by far the most accepted theory and is being used to explain the structure of the universe. Thus Newtonian physics which was based on solid bodies moving in an empty space, valid in our daily experience finds it extremely difficult to conceive a situation where these concepts are not applicable. Later, the concept of solid bodies was shattered in the realm of atomic physics.

In the theory describing the phenomenon on the basis of field theory, matter cannot be separated from curved space. Matter and space thus are inter-dependant and parts of an integrated whole. This is something radically different from Newtonian physics and leads to a notion of unity rather than fragmentation. It is this concept which raises new questions regarding the nature of reality. I salute to the genius of Einstein, who opened new horizons for understanding the relationship between objects and their surroundings.

One finds a striking parallelism between Einstein's concepts and the Eastern thought where time and space have been described as illusionary. This is illustrated by the following passage from *Madhyamik Karika*—the celebrated Buddhist text: 'It was taught by the Buddha, oh Monks, that ... the past, the future, the physical space ... and individuals are nothing but names, forms of thought, words of common usage, merely superficial realities.'

As this century comes to an end, one finds that human society has witnessed tremendous scientific progress and growth of unprecedented material affluence in many parts of the world on the one hand and a serious erosion of moral and spiritual values on the other. Mankind today is facing a multi-dimensional civilizational crisis which has engulfed almost all aspects of our lives—material and spiritual. All nations, whether rich or poor, are in fact nations with a troubled soul and are in search of a new paradigm. The question is how can science and spirituality reconcile and contribute to making this planet full of love and peace. Indian philosophy answered this question centuries ago and it would be my endeavour to place before the reader *how modern science is converging at the same point where ancient saints of India had reached and realized the Absolute Truth.*

The *Mundak Upanishad* poses a question 'What is that O' learned, knowing which all this universe of objective experience, both external and internal becomes known':

Kasminnu bhagavo vijnate sarvam
Idam vijnatam bhavati.

कस्मिन्नु भगवो विज्ञाते सर्वं इद विज्ञातं भवति

This is a wonderful question. The question has been deliberated on and answered in various ways in innumerable treatises by seers shining amongst them is Adishankaracharya, followed by great and learned Acharyas who interpreted the 'Truth' in many ways. This country, I think, is most fortunate to be the home of great and learned saints and in the past 150 years lived Sri Ramakrishna Paramahansa, Swami Vivekananda, Dayananda Saraswati, Swami Ramtirth, Sri Aurobindo, Ramana Maharishi, the Paramacharya of Kamakotipeetham and many others who not only experienced the Absolute Truth, but revealed it to others. To sum up—the understanding of *brahman* (absolute truth) only can lead one to complete knowledge and this is something all must strive for. But before proceeding further let the reader see what modern science has to say in this regard.

The scientific thought dominating the Western mind for the past three centuries evolved as a result of the works of

Galileo Galilei, Francis Bacon, Rene Descartes and Isaac Newton. The foundations of scientific rationalism were laid when Galileo for the first time combined experiments with mathematics. Bacon propounded a clear theory of inductive procedure and Descartes declared, 'all science is certain and evident knowledge'. Descartes also constructed an entirely new system in which events were mathematically described. His statement 'cogito ergo sum' (I exist because I think) resulted in a fragmented human personality with 'mind' separated from the 'body' and functioning as a controlling authority of the body. Descartes with his analytical skill portrayed the universe and all the objects which constituted it as automata. This world-view received further support from the works of Isaac Newton who developed his machines on the basis of the Cartesian view of the universe and made it the foundation of classical theoretical physics.

Newton showed the Western world that the universe was rationally comprehensible. In fact Newton's statement, 'I make no hypothesis (hypothesis non fingo) became the foundation for the rapid development of experimental sciences particularly in physics and ultimately resulted in some of the most wonderful scientific and technological discoveries—satellites, spacetravel! The industrial revolution could only be propelled because of his discovery of laws of motion which demonstrated that celestial bodies also obey the laws valid on this planet. Human mind could thus comprehend what till then was believed to be purely the domain of the divine. If Descartes developed the discipline of mathematical analysis and proved that nature could be pictured as a 'Great Machine', it was Isaac Newton who discovered the laws according to which the machine operated.

The Newtonian world-view was mechanistic, wholly deterministic, and governed by the principle of causality. In Newtonian physics the position of a moving object at any future time could be predicted provided one had the requisite information about its present position and the laws governing it's motion. Similarly, its past position at an earlier time could be retrodicted. Thus according to this concept the 'Great Machine' was operating like a pre-recorded tape where nothing could change and every event was predetermined right from the moment the machine came into existence. All this required the existence of an extraterrestrial agency to set the machine into motion in accordance with some divine laws.

Some of the philosophical consequences of Cartesian-Newtonian approach can be summed up as under:

(*a*) The universe is a giant machine and is governed by certain universal laws which can be discovered through experiments and rational understanding. By applying these laws one can extend the horizon of knowledge and can obtain further information about the nature of the universe.

(*b*) There is an external world which exists apart from this which can be observed and measured in an objective manner without producing any change in it. This external world is impersonal and the observer can strive for 'absolute objectivity'. That means man can observe the universe as a completely detached observer staying outside the universe

The theory of evolution developed as a result of research in geology, biology and the work of Pierre Laplace and of Immanuel Kant. In physics itself the discovery of the electro-magnetic phenomenon was not compatible with Newton's mechanistic model. The biologist started questioning the validity of the Cartesian concept that the universe came into being as a fully and perfectly constructed machine, they instead proposed an evolutionary paradigm in which the present day complex universe has evolved from simpler structures. Such a paradigm was also purely materialistic and accepted the view that the evolution of the external world can be uniquely and objectively observed by an independent observer. The spiritual elements were missing even in this approach. Despite these developments, science continued to quantify and measure, improve its experimental techniques and the Newtonian mechanics continued to be the basis of all physics till Einstein arrived on the scene.

After the successful experiments of Jagdish Chandra Bose (a person who has been forgotten by this country and almost by all scientists of the world), the distinction between animate and inanimate has disappeared. J.C. Bose started comparing the responses of non-living objects like metals and of animals. He discovered the fatigue of metals and then moved on from physics to physiology. On 10 May 1901 J.C. Bose demonstrated his experiments in England. Scientists saw with wonder the similar curves of muscles and metals, when they were responding to the effect of fatigue, stimulation, depression and poisonous drugs. He made the following remark after completion of his demonstration:

> I have shown you this evening autographic records of the history of stress and strain in the living and the non-living. How similar are the writings! So similar indeed that you cannot tell one apart from the other. Among such phenomena, how can we draw a line of demarcation and say, here the physical ends and there the physiological begins? Such absolute barriers do not exist ... It was when I came upon the mute witness of these records, and perceived on them one phase of pervading unity that bears within it all things—the mote that quivers in ripples of light, the teeming life upon our earth and the radiant sun that shines above us—it was then that I understood for the first time a little of that message proclaimed by my ancestors on the banks of the Ganges thirty centuries ago. 'They who see but one, in all the changing manifoldness of this universe, unto them belongs the eternal truth—unto no one else, unto no one else.'

Further, let's look at what James Lovelock says about the planetary ecosystem and what makes it tick. He says that: (a) the planetary eco-system is an entity where co-evolution is an accepted fact, (b) where highly integrative functioning is a key feature, and (c) where one cannot study a feature in isolation from others. The sum is, not only that everything is connected; but that one can understand the true nature of what does only by checking the relationships generally within its higher level contexts.

Examples can be multiplied to prove J.C. Bose's thesis. The Viennese biologist Raoul France, the American Clean Backster, the Japanese scientist Dr. Hashimoto and many others confirmed what J.C. Bose had demonstrated. The mechanistic world-view cannot explain this inter-connectedness of the organic and the inorganic.

Since the beginning of the twentieth century science has moved forward revising many concepts. Heisenberg's 'Principle of Uncertainty' has given a deeper insight in understanding the behaviour of sub-atomic particles. The strict cause and effect relationship breaks down in their domain. The determinism of the Newtonian model in the universe is replaced by indeterminism in the sub-atomic world. Theory moves from a deterministic world to an indeterministic one.

Science today is confronted with certain questions which it earlier considered to be outside its domain. But as a result of some of it's own discoveries intellectuals are asking, 'Is matter related to consciousness in any manner? If so, then what is the nature of this relationship?' It all began sometime around 1924–5 when Louis Victor de Broglie put forward the hypothesis of matter waves, i.e every particle of matter has waves associated with it and it is through the waves that one gets to know how the particles are behaving. This was a revolutionary concept. The thesis was sent to Einstein, who accepted it and Broglie got his degree at a very young age. Erwin Schrödinger—the father of the wave mechanical model—proposed his new equation which replaced Newton's equation of motion for a free particle in case of electron. Then came Werner Heisenberg with his 'Principle of Uncertainty' and stated that if one is certain about the position of a moving particle like an electron, one is uncertain about its momentum and vice-versa. The natural corollary of Heisenberg's 'Principle of Uncertainty' was: (a) That one cannot know exactly what a fundamental particle is—and fundamental particles are supposed to be the bricks of the universe, (b) that in the sub-atomic world the strict law of cause and effect breaks down, and (c) that the strict difference between an observer and the observed withers away. This is the most striking and revolutionary change which Heisenberg made, an observer participates with his observation, it is a participatory universe. One cannot remain objective and outside the universe and observe. The efforts to discover the ultimate reality through experiments are therefore meaningless, thus the classical concepts were no longer tenable in the sub-atomic world. The phenomena in this domain are statistically describable; but it is impossible to describe the behaviour of one particle with certainty.

Modern physics is now dealing with this new paradigm of quantum mechanics. The central question is: What is it that quantum mechanics describes. The answer generally accepted is known as the 'Copenhagen Interpretation'. This interpretation simply states that quantum mechanics is about correlations in one's experiences. It is about what will be observed under specified conditions. Einstein, however, opposed this till his last. His famous statement, that he did not 'believe in a dice playing God', expressed his disagreement with probabilistic interpretation of quantum mechanics.

The most startling consequence of the Copenhagen Interpretation was that the physicists under pressure of their own findings were forced to accept that a complete comprehension of reality lies beyond the capabilities of rational thought. It is significant that Einstein never agreed with this. However, the quantum mechanical paradigm unhesitatingly stated that the new physics was not based on 'absolute reality' but upon humans. The world 'out there' was inseparable from the observer 'in here'.

Einstein, who believed in causality could not accept Heisenberg's 'Principle of Uncertainty'. In order to disprove Einstein, Boris Podolsky and Nathun Rosen published a paper 'Can Quantum Mechanical Description of Physical Reality be Considered Complete?' The authors postulated that if the 'Principle of Uncertainty' was correct that causality does not hold good in the domain of the sub-atomic world, then it will lead to a strange paradox that two of the same kind of sub-atomic particles must somehow be simultaneously connected, even if they remain separated at enormous distances. How is it that these two particles communicate with each other instantaneously even at distances which electromagnetic waves take a few seconds to travel? Do they possess some kind of consciousness? Until 1936 no such phenomenon was known to exist, hence Einstein concluded that Heisenberg was wrong. But that the strange phenomenon did exist was proved by successful experiments in 1972 by David Bohm in London, by Clauser and Freedman in the USA and Alain Aspect's team in Paris in 1982. The impossibility of superluminary connection as propounded by the 'Theory of Relativity' is no longer valid. Hence 'an interconnectedness in events taking place at space-like distances is valid'.

J.S. Bell, a physicist at the European Organisation for Nuclear Research (CERN) gave a mathematical formulation of the EPR effect. The astounding implication of Bell's theorem is that at a deep and fundamental level the separate parts of the universe are connected in an intimate and immediate way. In 1975 Jack Sarfatti stated, 'not only superluminal connections exist, but they can be used in a controllable way to communicate messages'. Henry Strap in 1915 said, 'Bell's theorem is the most profound discovery of science'.

The most startling consequence of all these discoveries is that the Cartesian concept of reality as parts joined by local connections does not fit in the quantum mechanical paradigm. Henry Strap concludes by saying, 'the theorem of Bell proves, in effect, the profound truth that the world is either fundamentally lawless or fundamentally inseparable'. Bell's theorem implies that whatever happens in one part of the universe on a single entity has an effect which can be detected in any other part of the universe on a similar entity. Accordingly J.S. Bell's theorem has laid the foundation for the neodeterminism or superdeterminism in science. The initial conditions cannot be changed. The universe could not be anything but what it is.

David Bohm suggested that quantum physics demands a new order. 'Instead of starting with parts and showing how they work together, we start with the whole.' This is also in consonance with Bell's theorem. The separate parts of the universe are not separate parts. Says Bohm, 'Parts are seen to be in immediate connection, in which their dynamical relationship depends, in an irreducible way, on the state of the whole system (and indeed on that of broader systems in which they are contained, extending ultimately and in principle to the entire universe). Thus one is led to a new notion of unbroken wholeness which denied the classical ideas the analysability of the world into separately and independently existent parts.'

Explaining his hypothesis of apparently random sub-atomic phenomena, David Bohm says, 'Particles may appear in different places yet be connected in the implicate order. Particles may be discontiguous in space but they are contiguous in the implicate order.' Matter according to Bohm is a form of the implicate order as the vortex is the form of water—it is not reducible to smaller particles. Like 'matter' and everything else, particles are forms of the implicate order. The question which arises now is 'what is the implicate order?'

As Gary Zukov says, '*The implicate order* is the implicate order of *that—which—is*. This world-view is entirely different from what we are using in classical physics.' In the words of David Bohm: 'Description is totally incompatible with what we want to say'. Gary Zukov stated: 'Because of the deep rooted Greek notions in the Western mind, it is unable to comprehend this new paradigm. The Greeks believed that only *being* is. Therefore, *non-being* is not. Actually in the new paradigm *non-being* also is. Both *being* and *non-being* are "that—which—is". Everything even "emptiness" is "that—which—is".' In Bohm's physics, there is nothing which is not 'that—which—is'. Bohm's theories have striking parallelism in Eastern thought, in the Upanishadic statements.

The *Chandogya Upanishad* relates a dialogue between Svetaketu and his father. When Svetaketu returned home after studying the Vedas for 12 years, his father Uddalaka asked him, 'Svetaketu, have you asked for the knowledge by which we hear the unhearable, by which we perceive the unperceivable, by which we know the unknowable?' 'What is that knowledge?', asked Svetaketu. His father said, 'That knowledge is knowing that we know all'. And further explaining the father pronounced,

> In the beginning there was existence, One only without a second. Some say that in the beginning there was non-existence only, and that out of that the universe was born.
>
> But the question is how could existence be born of non-existence? In my opinion in the beginning there was existence alone—One only. He the One thought to himself: Let me be many, let me grow-forth. Thus out of himself he projected the universe, and having projected out of Himself, the Universe, He entered into every being. All that is has its *self* in him alone. Of all things He is the subtle essence. He is the truth. He is the *self*. And that, Svetaketu, THAT ART THOU.

In the *Brhad Aranyak Upanishad* the sage informs King Janaka about the true nature of *brahman*, '*brahman* can be apprehended only as knowledge itself, which is one with reality, inseparable from it. For He is beyond all proof, beyond all instruments of thought. The eternal *brahman* is pure, unborn, subtler than the subtlest, greater than the greatest. By the purified mind alone *brahman* is perceived. He who knows *brahman* to be the life of life, the eye of the eye, the ear of the ear, the mind of the mind,—he indeed comprehends fully the cause of the causes. In *brahman* there is no diversity. He who sees diversity goes from death to death.'

The quintessence of the Upanishadic thought is given in the following words:

Sarvamidam khalubrahman, ahambrahmasmi, tatvamasi

सर्वमिदं खलुब्रह्मं, अहंब्रंह्मास्मि, तत्वमसि

All this is *brahman*, I am *brahman*, so art thou.

Yatpinde tatbrahmande

यत्पिंडे तत्ब्रह्मांडे

That which is in the microcosm is also in the macrocosm.

Anoraniyan mahatomahiyan

अणोरणीयान, महतोमहीयान

Brahman is all pervading, it is subtler than the subtlest and larger than the largest.

So one notices that the language which many modern physicists are using is converging on the language of spiritual leaders. Thus J.C. Bose said: 'The vast abode of nature is built in many wings, each with it's own portal. The physicist, the chemist and the biologist come in by different doors each one to his own department of knowledge, and each comes to that which is his special domain unconnected with that of any other. Hence has arisen our present division of phenomena into the worlds of inorganic, vegetal and sentient. This philosophical attitude of mind may be denied. We must remember that all inquiries have as their goal the attainment of knowledge in its entirety.'

After 28 years of his first demonstration in 1901, J.C. Bose summed up his findings in 1929:

In many investigations on the action of forces on matter, I was amazed to find boundary lines vanishing and to discover points of contact emerging between the living and the non-living. My first work in the region of invisible lights made me realize how in the midst of a luminous ocean we stood almost blind. Just as in following light from the visible to the invisible our range of investigation transcends our physical sight, so also the problem of the great mystery of Life and Death is brought a little nearer solution, when, in the realm of the living, we pass from the voiced to the unvoiced.

That the modern mind and the ancient sages from India speak almost the same language is strikingly evident from the following two illustrations. In the words of Robert Oppenheimer: 'If we ask, for instance, whether the position of the electron remains the same, we must say "no"; if we ask whether the electrons' position changes with time we must say "no"; if we ask whether the electron is at rest we must say "no"; if we ask whether it is in motion we must say "no".' Now have a look at what the *Isa-Upanishad* says:

It moves. It moves not.
It is far, and it is near,
It is within all this,
And it is outside of all this.

Thousands of years before the dawn of civilization, as one understands it now, the learned had realized that it was the *brahman* from which the universe and all its beings arise; after arising in which they live; and at the dissolution into which they merge back.

yato va imani bhutani jayante, yena jatani jivanti,
yat prayanti-abhisamvishanṭi; tad vijijnasasva
tad brahmeti (*Taittiriya Upanishad*)

यतो वा इमानि भूतानि जायन्ते, येन जातानि जीवन्ति
यत् प्रयान्ति अभिसंविशान्ति, तद्विजिज्ञासस्व, तद्ब्रह्मेति

All one's inquiries have, therefore to be into the nature of *brahman* because at the level of knowledge it is abstract and hidden deep within every being and is not generally perceived; but it can be verily perceived with the help of intellect, sharp, subtle and one pointed by those who are adept in the perception of the subtle.

esha sarveshu bhuteshu gudho atma na prakashyate;
drishyate tu agryaya budhya sukshmaya sukshmadarshibhih.

एष सर्वेषु भूतेषु गूढ़ो आत्मा न प्रकाश्यते
दृष्ययते तु अग्रय्या बुद्धया सूक्ष्मया सूक्ष्मदर्शिभिः

The choice of words here is extremely sensitive. One indeed needs a very fine and sharp perception to see that the ***atman***, which is in every being. There are basic requirements too. Firstly devotion to truth, concentrated and concerted effort, perfect knowledge attained through constant control of extrovert desires, and practice of continence and other moral virtues.

satyena labhyah tapasa hi esha atma;
samyak-jnanena brahmcharyena nityam

(*Mundaka Upanishad*)

सत्येन लभ्यः तपसा हि एष आत्मा
सम्यक् ज्ञानेन ब्रह्मचर्येण नित्यम्

All through the philosophical literature one finds that to understand the *brahman* and *atman* (the self, the individual soul) simple knowledge is not enough. One has to have complete control on one's desires and practice of continence and other moral virtues.

At this point it may become evident that the *brahman* and *atman* is the same one infinite reality viewed from external and internal points of view. To understand the *atman* and *brahman* the mind is one's basic instrument and as Adishankaracharya has said:

The mind that has been well refined and rendered subtle by the study of the *shastras*, that is, scriptures and works dealing with spiritual science, and the teachings of a competent spiritual guide, and is well disciplined by the virtues of tranquillity of the mind and control of the senses, fortitude, withdrawal from extrovert desires, diligent conviction and constant concentrated application to realize the truth, is the instrument for the realization of the inner verities and the self (*atman*).

shastra, acharya-updesha, shama-dama-adibhih
susamskritam manah atma-darshane karanam

शास्त्र आचार्य-उपदेशः, शम-दम-आरिभिः
सुसंस्कृतं मनः आत्म-दर्शने करणम् (*शांकर गीताभाष्य*)

Vedanta, as summed up by Swami Mukhyananda holds,

That in the phenomenal state the universe is beginningless and endless, and evolves and involves cyclically in the cosmic consciousness by the force of the accumulated past impressions of the souls for their experience and action during many different lives to gradually progress towards perfection, to realize their true nature as the sat-chit-ananda. This can be illustrated empirically by our awaking and sleep cycles. When we go to sleep, all our stored knowledge and impressions become dormant, but retain the urge to come to the surface and remanifest. As a result, after waking up we resume our previous thoughts and activities and grow in knowledge and experience. This process goes on in life while our physical bodies are subject to constant changes and pass through childhood, youth, old age and death.

A question may be asked how the absolute becomes active and the whole universe acts accordingly. This energy has been explained as the *maya-shakti* or the energy of illusion. To quote:

Maya-shakti is the innate, eternal, indefinable and relativistic universe-manifesting power (anirvachaniya-shakti) of the trans-personal absolute *brahman*, which works spontaneously just as light shines spontaneously. *Maya* does not affect *brahman* just as the heat of the fire does not affect the fire but only others; that is, *maya* affects only its products. That is relativity. *Maya* projects the universe through its two fold characteristics of veiling-power (*avarana-shakti*) and projecting-power (*vikshepa-shakti*). Through the operations of these two powers of *maya*, on one hand it veils the reality or the real nature of *brahman/atman* which is existence-knowledge-bliss-absolute, and, on the other, simultaneously projects the *anatam* (non-self), the opposite or shadow of reality, in a distorted manner. This is so because of the disturbed condition of its three *gunas* (qualities), just as, for example, a concave or convex mirror distorts images. *Maya* projects the universe and its being on the substratum of *brahman/atman* in the relative plane, within the framework of space-time-causation (*desha-kala-nimitta*). This it does without affecting the absolute nature of *brahman/atman* just as a person is not affected by his distorted images projected by a concave or convex mirror. In the universe, which is within space-time-causation, individuation takes place, and the nature of *brahman* (*sat-chit-ananda*) appears dichotomized into relative existence and non-existence, namely birth and death, knowledge and ignorance, happiness and misery, etc.

Up to this point the discussion has been at a very subtle level. But it must be understood that spirituality and devotion to the *brahman* has to be at the level of the common man and therefore any discussion of Indian philosophical thought will not be complete without the devotional aspect of it. Knowledge, as has been discussed so far cannot be complete without devotion, and at some point of time knowledge and devotion converge supplementing each other to follow two different paths to reach the same goal namely the realization of the *self*.

As I said earlier, the path of knowledge being difficult, it is restricted to a few. But the path of devotion (*bhakti*) is available to one and all who wish to experience the divine love. Shankaracharya had established the *adwait-vedanta* or principle of non-duality and the following masters enriched this knowledge through their own treatises. But Shankaracharya was also a great devotee and his *Saundarya-lahari* and other hymns are sung by millions everyday. Thus pure knowledge has always followed the path of devotion to bring the unmanifest to the level of the manifest. In devotion the abstract becomes concrete.

Bhakti or devotion is supposed to have two sons: *gyan* (knowledge) and *vairagya* (renunciation). Knowledge without renunciation has no meaning. When knowledge is coupled with renunciation, devotion takes form.

In this context I would refer to two great works on devotion in Indian philosophy. These are the *Narad-bhakti Sutra* and the *Bhagwat Mahapuran*. I will not go into the details but only touch upon the fundamentals. The *Narad* says, 'Devotion is indeed in the nature of supreme love in the divine and the supreme love of God is in the nature of immortality.'

Param premroopa, amrit swaroopa cha

परम प्रेमरूपा, अमृत स्वरूपा च

Having attained devotion the being becomes perfect or becomes a *siddha* and what is the state one achieves after that? 'Having achieved devotion to God, one becomes devoid of desires. He does not grieve, he is not jealous, he does not rejoice and he is not enthusiastic for self-interest. In this state, the devotee becomes intoxicated in divine love, is in complete peace with himself, becomes silent and becomes united in the self. Such is the nature of devotion.'

The *Narad-bhakti Sutra*, however, says that the nature of devotion is beyond description. It is like the taste of sugar enjoyed by the dumb. Devotion takes one to the state of supreme bliss and highest spiritual experience which is called '*bhooma*'. This term has been used in several Upanishads as the highest plane of fulfilment. This has been defined in the *Chandogya Upanishad*, 'In this state you see nothing else, you hear nothing else, nothing else you know and feel. That state is called *bhooma*—the all pervading, the infinite, the eternal, the immutable *brahman*. There where you see other things, where you feel and understand other things, that is finite and limited. That which is infinite *bhooma*, is immortal.'

यत्र नान्यतपश्यति नान्यच्छृणोति
नान्यिद्वजानाति स भूमा अथ यत्रान्यत्पश्यति
अन्यच्छृणोति अन्यद्विजानाति तदल्पं यो वै भूमा
तदमृतमथ यदल्पं तन्मर्त्यम्।

Yatra nanyatpashyati nanyatchhrinoti
nanyadvijanati sa bhooma, atha yatraanyatpashyati
anyatchhrinoti anyadvijanati tadalmpam yo vai bhooma
tadamritamath ydalpm tanmartyam.

Most people are propelled towards worldly happiness, because their experience is limited to the things that they have in some way come across. The divine experience does not come easily and, therefore, it is an abstract happiness. Who would be interested in abstract happiness if one is able to experience the worldly one? But this enquiry to understand the nature of the divine opens a vista for another kind of experience and one starts thinking, meditating and contemplating on the ultimate. This attention to *brahman* has to be complete and is described as oil being poured from one container to another is uninterrupted and smooth, so should be concentration (*taidharawat*). Once this happens, the divine enters one's heart.

The understanding of the divine makes one's heart pure. A strange happiness which is unrelated to one's daily existence and worldly happiness is felt. As it happens, one is tied to this world with an unreal knot. Once the divine ray enters one's heart one is able to see that knot. As soon as it is seen, the unreality of it can be understood. And then, as the *Bhagavat Gita* says, this knot is untied. In other words, all the doubts, with which one's thinking is conditioned, completely disappear. The cycle of *karma* (past actions) breaks and one is face to face with this truth. Thereafter, nothing more remains to be understood. The person becomes one with himself. All action ceases and there is nothing save peace and love. One is within the Lord and the Lord is within one. Not only that the Lord follows the devotees. What a beautiful concept! Krishna says to his dear friend Uddhav in the *Bhagavat Gita*:

> O Uddhav! loving devotees like you are the dearest of all to me; even my son, Brahma the creator, Shankara, my real brother, Balram and my consort Lakshmi, even my *atman*, my own self are not so dear to me as my devotees are. I constantly follow such devotees of mine who have no desire, and are only absorbed in meditating and contemplating upon me, discarding every thought of the world, who are free from attachment and aversion and have compassion for everyone, so that the dust raised by their feet may fall upon me and I may be sanctified thereby. Those who have renounced all acquisitions and possessions, those who do not have any attachment, ego or vanity even in relation to their body, etc., whose mind is imbued with exclusive love for me, who have reached a state of tranquillity by renouncing all worldly desires, who because of their generous and exalted naturet reat everybody with compassion and love and whose minds are totally unaffected by any desire whatsoever, experience the supreme bliss which cannot be experienced by any other person as this supreme bliss can be experienced only by renouncing all expectations and desires.

It is indeed an elating experience that the Lord follows the devotee for the dust of his feet in order to sanctify Himself.

I started with the words from the *Mundak Upanishad*, 'What is that O' learned, knowing which all this universe of objective experience, both external and internal becomes known.' The answer to this is that the knowledge of absolute truth of the *brahman* can only lead one to love and peace and when this happens the *Mundak Upanishad* declares 'when the *atman* is realized as the basic reality in the noumenal and phenomenal existence, all the complexities within the heart (mind) are resolved and all the doubts are rent asunder...'.

Bhidyate hridaya-granthih, chhidyante sarva-samshayah,
kshiyante chaasya karmani tasmin drishte paravare.

भिद्यते हृदयग्रन्थिः छिद्यन्ते सर्वसंशयाः ।
क्षीयन्ते चास्य कर्माणि तस्मिन्दृष्टे परावरे ।।

Scientists today are discussing questions like 'reality', 'being', 'non-being', etc., in almost the same language which the Upanishads and other schools of Indian philosophy have done. Piet Hut in one his recent papers (1995) while dealing with these issues has observed the following:

Science that does not have any ethical implication can be useful, but cannot claim in any way to describe all of reality since clearly some form of ethics is part of our world of experience. On the other hand, ethics as a set of arbitrary commands, either ascribed to a superhuman source or to a biologically useful set of general rules, is not satisfactory either.

What is this world humans are living in and who are they? In order to come to terms with such questions, one must switch from an inquiry as to 'what' to a more revealing inquiry as to 'how'. How does this whole world arise in the way it does, in the way it appears to one? And how does the notion of who one is arise in that same experience in which the world appears as well?

Asking such questions, one can find a tentative answer to what it means to say that something 'is'. There 'is' a cup, there 'is' joy, there is form and function and value. Whatever appears, it has to make some. form of sense to one, in order to qualify as something that 'is'. Even utter chaos presents a form of sense (namely: chaos, nonsense). So, for one, 'what is' is the direct result of identifications one has made.

At any time one can view anything in its 'being' aspect as the role that is being played, as that 'what it is'. But one can equally view it in its 'non-being' aspect, in its aspect of openness or emptiness. From the point of view of the play, the player underneath the role being played is simply not there. In a drama, there 'is' a king. The actor 'is not' within the rules of the play. Within the play the actor steps aside, disappears, to let the king show through. But when one steps outside the play the king has vanished, has completely lost his base, his foundation of existence. One then sees that, at bottom, the 'king' has been an empty notion all along. something being played but not ultimately 'real' in any sense.

It would seem that, instead, the actor is the real person, rather than the king. But if the actor unexpectedly gets fired, soon after the play? Then the next layer drops. The actor disappears as well, and a jobless person appears instead. Is there a core that remains unchanged? Flesh and bones, or molecules, or a life history constructed as a vast web of connected past events and relationship between events? Or are any and all of these, in turn, the results of further attempts at role-playing and play interpretations? I suspect the latter. Let me try to sketch what that may mean.

Stated in the most radical way, each subject or object, human or physical object or abstract idea, or whatever, is playing a role. And one identifies as playing a role is itself playing a role. All are part of a great drama of role playing, with roles within roles within roles—without anybody or anything 'home' underneath; without any stable or final foundation to bolt things down upon.

There are interesting parallels with mysticism though. If a scientist looks at the claims of a traditional religion, he or she is likely to be rather sceptical at the seemingly arbitrary boundaries that are acknowledged by the true believers of that religion. To take one example, the believer may claim that a particular temple ground is 'holy', and that there is a clear distinction in sacredness between what lies within the perimeter of the temple area and what lies outside. The scientist may object that the molecules inside the temple are the same as those outside, and that no scientific analysis is likely to yield any measurable distinction that could justify the presence of a definitive limit, separating the sacred from the non-sacred. 'Such a scientific attitude would be very reductionist and would probably not convince the believer. The scientist, in turn, might well be ready to concede a contextual value to the notion of sacredness. But it is interesting to take up the reductionist conclusion and turn the tables. The first naive interpretation of the absence of a distinction between temple and non-temple would suggest that nothing is sacred. But this is only one way to react to the dropping away of a limit. There is an alternative: one may equally well conclude that everything is sacred. Rather than limiting appreciation to a particular spot, one can follow the examples of mystics of all ages, who have never tired telling mankind that there is no such thing as ordinary, finite, non-sacred things and events. Anything can be viewed in its proper aspect, as an open gateway to a boundless reality. It is here that science and mysticism meet, in an outlook that is literally limitless.'

Science and Ethics

STEPHAN F. VON WELCK

The topic 'Science and Ethics'—or to be more precise 'Science, Research, Technology and Ethics'—is not a new one. It has been discussed within the scientific community for many years, if not decades. During the 1990s, the debate on the role of ethics in science has broadened considerably, due to new and far-reaching discoveries and inventions especially in biosciences and biotechnology, such as genetic modifications of plants and animals, research with human embryos or the clon-ing of animals with the prospect of human cloning. As a consequence, the debate now involves not only scientists but also philosophers, lawyers, the medical profession, religious leaders and politicians. During last year's World Conference on Science at Budapest, organized by UNESCO and the International Council of Scientific Unions (ICSU) and attended by representatives of more than 180 countries and numerous international organizations, the interaction between science and ethics has for the first time been discussed under such a broad heading in a worldwide policy forum, thus becoming a topic of priority on the agenda of UN science policy conferences. What is still rudimentary, however, is an informed discussion of the subject outside of expert meetings, that is to say a scholarly and profound interaction between the experts and the interested and concerned public on the role of ethics in science and the ethics of science. For this reason I thought that the topic 'Science and Ethics' could well be the subject of the last Bose-Einstein Lecture on Science, Technology and Environment.

*Twelfth Bose-Einstein Lecture, 9 March 2000, India International Centre, New Delhi.

II

The development of science, research and technology during the twentieth century is a phenomenon, which has profoundly changed the globe, the society and day to day life. Most of these changes are certainly of great benefit to humankind. Due to scientific progress, life expectancy has increased considerably, cures hitherto unknown have been discovered for many diseases, agricultural output has risen significantly in many parts of the world, including India, to meet shortfalls and the demands of a growing population. Technological developments based on scientific advances and the use of new materials have produced a so far unknown range of industrial products and processes which facilitate considerably one's daily life. New methods of communication, information handling and computation have brought unprecedented opportunities for society at large. The steadily growing knowledge on the origins, functions and evolution of the universe and of life on the planet Earth has provided mankind with conceptual as well as practical approaches that profoundly influence life.

This unprecedented penetration of nearly every aspect of one's life by scientific knowledge and innovation is likely to continue during the twenty-first century, and most probably at an ever increasing speed. Of course, it is difficult to predict the future development of science, technology and innovation and their impact on one's life and that of generations to come. But it seems fair to say that one will have to cope with a proliferation of knowledge much faster than up to now and that due to scientific progress the world will look very different in 50 years time from the one at the beginning of the year 2000. The way one lives will dramatically change. This is true for all countries, those which are called developed today and those which are still in the process of development, and most certainly for a country like India that has such a wealth of scientific talent and capacities and at the same time such tremendous problems waiting to be solved by translating the results of modern science and technology into practical applications.

III

While science, research and technology have changed the social and natural environment dramatically and will continue to do so, the risks and challenges have increased as well, on both ends of the scientific and technological process.

As the worldwide demand for research and new technologies continues to grow while available funds get shorter, competition amongst research institutions and scientists has increased and new alliances have been formed. Alliances with economic powers like industry have become commonplace and the laws of the marketplace sometimes overshadow the more traditional values and norms of scientific research like objectivity, openness or responsibility. Many prestigious research institutions, especially universities, struggle with their new role as actors in a national or even global knowledge market. Some are deeply worried by what they perceive as an ever increasing race for patents, outside funding or excellence ratings. The competition on markets is reflected inside the scientific community itself, as competing groups of experts race towards the first marketable breakthrough, triggered not so much by their interest in gaining new knowledge but by large economic incentives. In addition, more and more knowledge is protected as private property and has to be paid for according to market forces. It provides privileged access to advanced technologies and thus competitive advantages in the striving for larger markets, greater profits and added economic power.

In order to limit the risk of science becoming too dependent on economic factors and market forces, the new alliance of science with economic power (as, by the way, the old and still flourishing alliance of science with military power) must be balanced with science exercised in the interest of civil society. Obviously, not all goods a society needs are goods for the market or are useful for defence purposes. Some are cultural, spiritual or aesthetic, some provide new intellectual dimensions or new insights into the origins of the universe, such as those produced by S.N. Bose and Albert Einstein. If science neglects these areas, it risks to be seen by the public as an obedient servant of purely economic or military interests. It risks to loose trust. And in a democracy, nothing can be more damaging for the long-term prospects of science and technology than public lack of trust, and especially if the young generation doubts scientists and scientific institutions.

IV

On the other end of the 'scientific-technological production line', namely the output, even more dramatic risks and challenges have become obvious. People have realized that science, research and technology not only 'produce' improvements in the quality of life, but also results which may very well turn out to be detrimental to the society in the long-run, or even cause man-made disasters such as nuclear wars, natural catastrophes or human diseases with consequences mankind has so far not even imagined. This dual potential of science, research and technology to do both, improve and harm, is the very subject of this article.

Albert Einstein, one of the two men of genius after whom these lectures are named, was at a later stage of his life very aware of this ambivalent character of science and technology, including his own revolutionary insights and theories. Knowing that his 'General Theory of Relativity' had an undeniable impact on the understanding of nuclear fission and consequently on the invention of the atomic bomb, he strongly campaigned in his later years for a ban on nuclear weaponry.[1]

With the same objective, 18 German scientists, among them several Nobel laureats like Max Born, Otto Hahn, Werner

[1] Albrecht Fölsing, *Albert Einstein*, New York, 1997, p. 737

Heisenberg, Max von Laue and Wolfgang Pauli, raised their voice against the misuse of science and research for military purposes. In their famous 'Göttingen Declaration' of 12 April 1957, they urged the Government of the Federal Republic of Germany to abstain from any plans to introduce nuclear weapons into the German army, and declared that none of them would in any way participate in the production, testing or use of nuclear weapons.[2] Satyendra Nath Bose, too, warned of the dangerous implications of science and research. In his talk at the Tagore centenary celebrations at University College of Science, Calcutta in 1961, after having criticized the sometimes insensible attitude of scientists concerning the long-term consequences of their activities, he said, 'If we are not cautious, if we do not change our attitude, this earth may cease to exist.'[3]

One of the first non-scientists who devoted a large part of his intellectual work to ethical issues arising out of science and technology and their applications was German born philosopher Hans Jonas, who taught from 1955 to 1976 at the New School of Social Research in New York. His publications on relevant problems like *Philosophical Reflections on Experiments with Human Subjects* (1969), *Towards a Philosophy of Technology* (1979), *Technology as a Subject of Ethics* (1982) or *Science, Ethics and Politics* (1983) have strongly influenced the debate on science and ethics at least in the Western world.[4]

His main thesis is that science and technology are not just a means to acquire new knowledge and new products, but a very special way to exercise human power, and should therefore be subject to a special set of ethical rules, which are different from those being applied to other human activities.

[2] *Frankfurter Allgemeine Zeitung*, 13 April 1957, p. 1.

[3] S.N. Bose, 'A Scientist's Apologia', in S.N. Bose, *The Man and His Work*, Vol. II, S.N. Bose National Centre for Basic Sciences, Calcutta, 1994, p. 222.

[4] Hans Jonas, *Das Prinzip Verantwortung*, Frankfurt a.M., 1979; Hans Jonas, *Technik, Medizin und Ethik: Zur Praxis des Prinzips Verantwortung*, Frankfurt a.M., 1987, with references to publications in English, pp. 323–4.

Without fully sharing this rather generalized view, Hans Jonas' plea for a 'special responsibility' of scientists, researchers and engineers who work at the forefront of modern science and technology, is a demand to which I can certainly subscribe to.

V

Let me share this view in some more detail by examining the discipline of modern science and technology which might have the most far-reaching influence on society during the decades to come, namely bioscience and biotechnology.

During the first half of the twentieth century, science opened the door for understanding and changing the secrets of matter through nuclear physics. Now within only a few decades, science has opened the door for understanding and altering the secrets of living nature, of life itself, through biology. This is no more the old descriptive biology which one used to learn during one's schooltime. It is a fundamentally new type of biology,* namely the analytical and constructive science which today is called molecular biology or bioscience. This modern biology has found an ever increasing spectrum of applications in biotechnology, medicine and genetic engineering.

What makes bioscience and its application so outstanding if not to say revolutionary, is its unprecedented impact on the understanding of and intervention in living nature, in life itself. It enables scientists and genetic engineers not only to alter nature as it surrounds mankind, but to change human's own nature. This is new in the history of science and technology, it gives science a new dimension, and provides human beings—to say it in the words of Hans Jonas—with a new tool to exercise power.

What has already been achieved in this field of bioscience, biotechnology and genetic engineering?

In plant biology, for example, a variety of genetically modified foods has been produced by transferring defined genes into traditional food crops, resulting in improvements such as higher yields, better quality including nutritional quality and storage properties, as well as better disease and pest resistance of plants. If one wants to meet the increased demand for food caused by population growth in many countries of the world, and if one wants to replace chemical farming with its detrimental effects on soil and food, one may have no other choice but to continue using these new means of plant biology which pave the way for the socalled second green revolution.

In health and medicine, new techniques and therapeutics using recent advances in bioscience are already in practice or undergoing trials. In vitro fertilization and other methods of assisted reproductive technologies are standard practice in many countries. Genetic diagnostics and genetic therapy have been introduced in modern medicine. Transplantation of animal tissues or even complete organs into humans may soon be possible, thus overcoming the problem of shortage of human donors for transplant surgery. Recent advances in cell biology are even more spectacular. The possibility of maintaining pluripotent embryonic stem cells of human beings in culture opens a novel dimension of medical research. For the first time it has become feasible to study in vitro the so far largely unknown highly complex processes of human tissue differentiation and of organ formation. Research with such human cells may result in the development of cell transplantation therapies for disorders such as Alzheimer's disease or cardiovascular diseases and cancer, for which other therapeutical means are not yet available or still unsatisfactory. It could also open up the possibility to prepare, from the body cell of a patient and a mature fertilizable egg cell, embryonic stem cells containing the patient's own genetic material, a technique which might finally lead to the cloning of human beings. Another field of highly sophisticated health research attracting a lot of interest is genome research. It aims at the decoding of the complete genetic make-up of human chromosomes, thus revealing vital information on several hereditary diseases such as schizophrenia or deafness.

VI

The explosion of bioscience and genetic engineering techniques during the last decade is, however, a two-sided coin. It has created breaches in the boundaries which used to mark the limits of human intervention. The resistance of nature, which has imposed these boundaries on mankind over the millennia, has so far given mankind a stable foundation. If this resistance now disappears across broad areas, mankind as a whole may be at risk. The questions raised by these developments are indeed of a fundamental nature. They are questions which in their final consequence touch upon the relationship between God as creator and human beings as his creatures.

Is it permitted to change the genetic make-up of plants to make them more useful for human ends without knowing their possible long-term impact on consumer health and on the planet's ecological balance? Is it acceptable to 'make' children in test tubes, thus intervening in the natural selection process? What about the transplantation of animal organs to humans, thus trespassing the line between animals and human beings? Should one allow scientists or genetic engineers to determine the character of individual beings by changing the genetic make-up with the aim to avoid for example criminal tendencies inherited through so-called crime genes? Is the world heading, in the final consequence, for a society based on prenatal selection between 'good' and 'bad' genes, a society of human beings who are 'genetically bred', formed not only by education or the environment in which they live, but foremost by 'breeding programmes' and 'breeding strategies'?

These are questions of disturbing, if not frightening character. They have, formulated in these provoking terms only recently, raised a heated public debate in my country, triggered by a widely published and commented lecture of the controversial philosopher Peter Sloterdijk on *Regeln für den*

Menschenpark (rules for the man-park), in which he calls for establishing a 'Codex for Anthropotechniques'.[5]

VII

Even the most ardent critics of Peter Sloterdijk agree that there is indeed a need for limitations, that not everything that can be done, may be done. They agree that the new science, especially bioscience and its applications, can no longer stick to the traditional rule of ethical neutrality which has governed science during past centuries. It intervenes too deeply, too irreversibly into nature and with too many long-term consequences for society. Modern science has to accept being judged on the basis of ethical criteria as well, namely of good or evil. The pertinent question in this regard is, however, where to draw the line between good and evil science, between what is still permitted and what should be prohibited on the basis of ethical and finally legal norms.

This brings one to the principle of freedom of science and research, which has been the basis for the unprecedented advances of science, research and technology during the last century, and which is guaranteed by the constitutional law of many countries, including Germany. The vast majority of constitutional lawyers today agree that the freedom of science and research is not unlimited, even if the constitution—as for example the Basic Law of Germany—does not provide for restrictions in explicit terms. It may itself be subject to limitations by other constitutional rights, such as the fundamental right to protect human dignity or human life and health. Any restrictions on the basis of ethical values however, have to take into account that because of the plurality of moral convictions, it is most difficult to agree on common ethical principles which could serve as a basis for legal regulations limiting the right to freedom for science and research.

[5] Peter Sloterdijk, *Regeln für den Menschenpark: Ein Antwortschreiben zum Brief über den Humanismus*, Frankfurt, 1999, p. 17.

German legislators had to face this dilemma when drafting the Embryo Protection Law of 1990, which aims at safeguarding human dignity and the protection of life from its inception.

At the international level it is even more difficult to achieve a consensus across national borders and across the related differences in ethical concepts. The document which first confronted the German public with the whole weight of this problem was the draft of a *Convention on Human Rights in Biomedicine*[6] submitted by the Council of Europe in 1991. After a heated public debate and a number of changes, this draft was finally approved by a majority of member countries of the Council of Europe in November 1996 and has become binding international law through ratification by five countries in December 1999. Germany, however, has so far neither signed nor ratified this convention, because of the government's opinion that the provisions of the convention do not go far enough in their efforts to protect human rights and the dignity of human beings. The German public is particularly sensitive regarding any intervention in human life and dignity, triggered not only, but mainly by the memory of criminal medical experiments during the Nazi regime. This example shows clearly the problem of consensus building on ethics and law relating to the boundaries of bioscience, modern medicine and genetic engineering.

Similar difficulties arise with regard to multilateral arrangements such as the UNESCO's *Universal Declaration on the Human Genome and Human Rights*, which was adopted unanimously by the General Conference of UNESCO in November 1997. This declaration also tries to strike a balance between the fundamental freedom of science and research on the one hand, and the protection of human rights

[6] See *Convention for the Protection of Human Rights and Dignity of the Human Being with Regard to the Application of Biology and Medicine*. For more information on this debate see: Ludger Honnefelder, 'Science and Ethics: the Concept of Human Rights as the Basis of a European Concensus', in *Bildung und Wissenschaft* 2/1998, pp. 3–13.

and dignity on the other hand. However, since the restrictions on science and research are only in the legal form of a declaration and not of binding international law, the difficulties in arriving at a consensus among member countries have not been as dramatic as it was the case with the European Convention.

After all, these legal norms, whether in the form of national laws, international conventions or universal declarations, are subject to interpretation by individual scientists, engineers or the medical profession as well as by those whose main intention is to safeguard the dignity of human beings against the risks and dangers of modern science and technology. In the last resort, it is up to the individual scientist, engineer or medical doctor to make his/her personal assessment as to the ethical justification of applying the means and instruments which modern science puts at their disposal. It is his/her personal responsibility to draw the thin line between what is ethically still justified and what should not be done. No law or convention can liberate either from this personal obligation.

What governments and the society at large can do to share this burden, is to encourage the teaching of ethics in scientific education in order to strengthen ethical awareness among the individual scientists, to establish independent national bodies for ethics in science, to formulate codes and guidelines, and to encourage an open and informed dialogue between the scientific community and the general public on this topic. This will help increase ethical awareness within the community and control fears of the public about its future. For it is up to us to decide whether we will control science and technology or whether science and technology will control us.

Plate 5: Rabindranath Tagore exchanged ideas with Albert Einstein several times during his journeys to Europe. He visited Einstein at his house in Caputh near Berlin in 1931.
Courtesy: Rabindra Bhavan, Shantiniketan

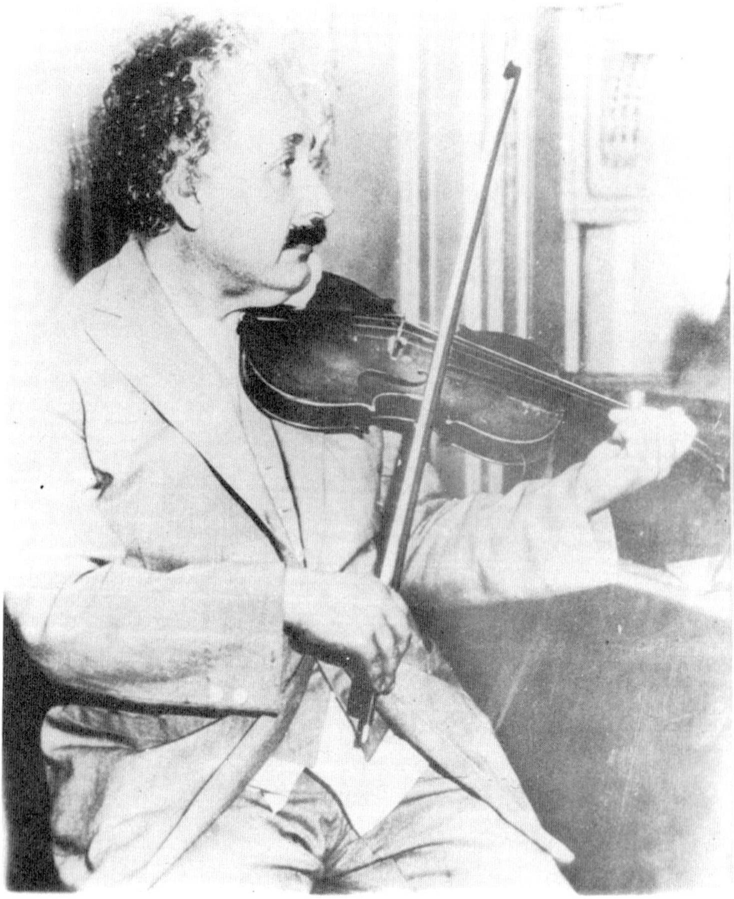

Plate 6: Albert Einstein and S.N. Bose shared a talent and love for music: Einstein played the violin since his childhood.
Courtesy: Birla Industrial and Technological Museum, Calcutta

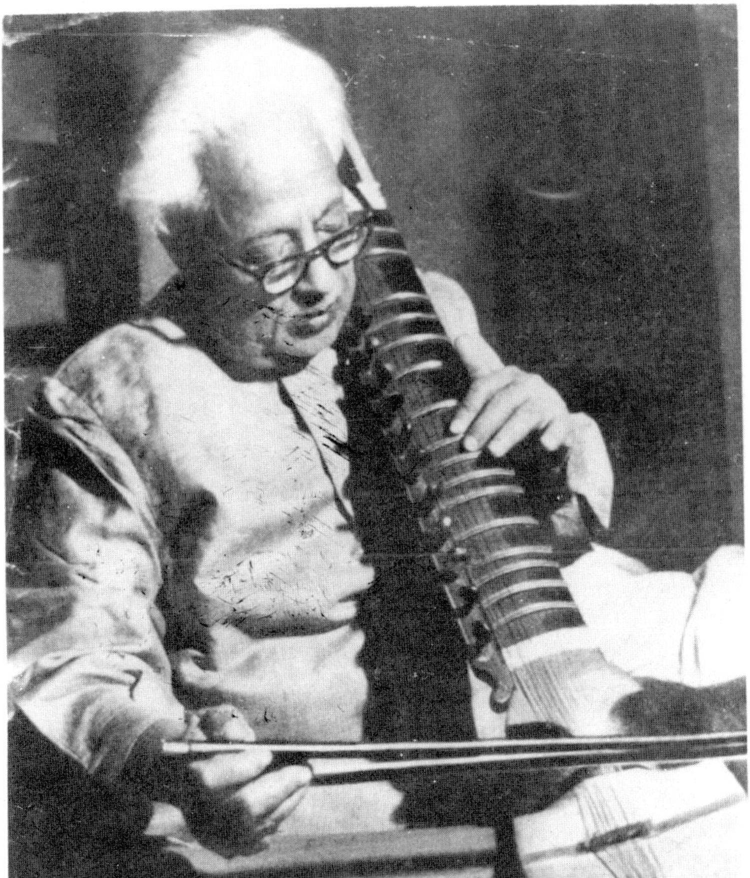

Plate 7: S.N. Bose was a master on the esraj.
Courtesy: Birla Industrial and Technological Museum, Calcutta

Plate 8: Get together after the 8th Bose-Einstein Lecture on 'Science and Religion'. The speaker Prof. M.M. Joshi, Minister for Human Resource Development, Prof. A.M. Khusro, Dagmar Bernstorff, the consulting editor.
Photo: S.K.Sharma

Science, Technology, Innovation: Their Impact on Economic and Political Power

R.A. MASHELKAR

Introduction

The Bose-Einstein intellectual bond was both creative and durable. The first ever English translation of Einstein's paper on relativity, originally published in German, was made by S.N. Bose and his distinguished colleague Meghnad Saha and printed in Calcutta. The word 'boson' to describe elementary particles that obey the Bose-Einstein statistics is now a part of scientific language. The recent Nobel prize winning work has put the experimental seal on this inspired theoretical work.

I will outline some of my perceptions of the way science, technology and innovation are going to determine the economic and political power of nations. The rapid paradigm shifts that are taking place in the world as it moves from superpower bipolarity to multipolarity, as industrial capitalism gives way to green capitalism and digital capitalism, as information technology creates netizens out of citizens, as aspirations of the poor get fuelled by the increasingly easier access to information, as the nations move from 'independence' to 'interdependence', as national boundaries become notional, and as the concept of global citizenship evolves, we will be full of new paradigms and new paradoxes, there is no doubt that the rapid advance of science and technology will directly fuel many of these.

*Eleventh Bose-Einstein Lecture, 13 December 1999, India International Centre, New Delhi.

Science, technology and innovation: the changing paths

The paths in science, technology, and innovation have undergone a remarkable change. In the pre-World War II scenario, the only driving force was the search for knowledge. No substantial government funding was available. Scientific research was done by a handful of small enterprises. There were very few industrial scientific research laboratories and very few industries that supported scientific research. Little interaction existed between academics and industry. The post-World War II scenario saw the parenthood of research change. On the one hand, research was entirely driven by the search for new knowledge and on the other hand by economics, defence and health. The government became a major funder for research. The science-based industries, such as biotechnology and information technology grew. There was also a growth of industrial scientific research. One could walk into a Bell lab and see Nobel laureates. There was an intense industry-academia interaction.

Then came the post cold war era. The defence base for science started declining. Economic growth and health became the prime drivers for generation of new knowledge. Industrial basic research started vanishing. Innovation driven by defence spending and its subsequent diffusion into society through technology 'spin-offs' came in for a major revision.

Advances in knowledge during the post cold war era also brought up new issues. Genetic engineering and the associated reproductive technologies on plants, animals and humans brought forthethical issues calling for greater regulation by involving social scientists and environmentalists. The process of globalization, privatization and corporatization of research changed the dynamics of the creation of knowledge. Issues of Intellectual Property Rights (IPRs) and proprietary information and knowledge have begun to give rise to new debates on public good *versus* private profit. New models of the innovation chain and new paradigms of the science-economy and science-society contracts have begun to emerge.

Science and technology models: the paradigm shifts

Some of the old models of science, technology and development are giving rise to new models. Francis Bacon, in his book *Advancement of Learning*, published in 1605 had established a linear model of scientific discoveries leading to technology development, which in turn, led to economic development. This model is certainly not valid today. Technology is not always the offspring of science. Quite often, technology precedes science. The steam engine came before the laws of thermodynamics were understood. A major part of new technologies indeed evolve from already existing science and technology. Many advances and innovations in technology are essentially incremental improvements in existing technologies. A technology can give rise to new technologies, the so called spin-off technologies. It is not only that new science gives rise to new technology, but the reverse is also true: New technology gives us a new science.

The other important model was put forward by Vannevar Bush in 1954 in his famous document called *Science: The Endless Frontier*. It remained the key framework of the science policy in the West and was also followed by many developing countries. The 'Bush Model' was a linear one, which assumed that 'developed science' will eventually spread into society as useful products.

This linear chain of innovation disregarded the socio-economic, organizational, cultural and other contexts including such factors as the character of the marketplace, both domestic and global. Hence, pure science was valued above applied science or technology and the connection of science to its societal context was devalued. As a recent historical evaluation of the Bush report observed, 'Societal benefits (in the Bush model) result not in spite of isolation from the broader environment, but rather because of isolation, as autonomy is a necessary element of the scientist's ecology.'

Subsequent events have shown that the innovation chain is far more complex than this simplistic linear, unidirectional paradigm of science push based development. Even in the context of the West, particularly with the emergence of Japan as a technological power, a mix of supply-driven and demand-driven science policy doctrines, for example the OECD reports and the Foresight initiatives modelled on the Japanese experience, began to evolve. The boundaries between basic, applied and socially mandated sectors were becoming increasingly blurred and a hybridization of scientific communities, which divided tasks between laboratory and industry through contract research and consultancy, was taking place.

In the post cold war era, the second element of the old contract, namely innovation driven by defence spending and its subsequent diffusion into society through technology 'spin-offs' has also come in for a major revision in countries like the USA, where defence spending has led to investment in R&D.

In a global context, therefore, a major science and technology issue of discussion is the 'conversion of military to civil technologies' and the new concept of 'dual use technologies', rather than spin-offs. Arguments of the past against a defence-led mode of technology innovation have now given way to economic arguments, based on the successes of the industrial policies of Japan and Europe led by explicitly civilian technology.

For example, the aim of the new US strategy is to redirect the massive US defence investment so that it will become more effective and more supportive for a broader industrial base. As the 1993 US technology policy states: 'We cannot rely on the serendipitous application of defence technology to the private sector. We must aim directly at these new challenges and focus our effort on the new opportunities before us, recognising that the government can play a key role in helping private firms develop and profit from innovations.'

The US Technology Reinvestment Project of 1993 is a strong indicator of how dual-use technology development, instead of the notion of spin-offs, is intended to be pursued in cooperation with industry as a technology diffusion policy approach. This programme is aimed at stimulating a transition of defence technologies to a growing, integrated national industrial capacity.

Let the focus now turn on the perspectives of the developing world. The process of globalization, dictated by the new global trade regimes under the WTO, rapid advent and influence of information technology on all spheres of life, issues linked to intellectual property rights, the growing awareness of the value of the rich traditional and civilizational knowledge systems are becoming crucial in the new paradigm.

The IPR issues: the North-South philosophical divide

I was attending a meeting of the Third World Academy of Sciences in Trieste recently. Frederico Mayor, the Director General of UNESCO, made an interesting statement. He said, 'Knowledge flows from North to South and wisdom flows from South to North'. I remember making a small correction; 'knowledge' may flow from North to South but 'usable knowledge' does not flow that easily from North to South, since usable knowledge has the potential to create wealth. No country or no corporation gives a competitive advantage to another, excepting at a price. India itself has realized this in the post liberalization era.

The basic conflict between the North and the South can be simply summarized by the fact that whereas (usable) knowledge does not flow that easily from the North to the South, wisdom flows from South to North rather easily. One must understand what is meant by 'wisdom' here. Many societies in the developing world have nurtured and refined systems of knowledge of their own, relating to such diverse domains as geology, ecology, botany, agriculture, physiology and health. One is now seeing the emergence of terms such as 'parallel', 'indigenous' and 'civilizational' knowledge systems. In some sense, the flow of 'wisdom' from South to North encompasses all these and more. However, the issues that get linked to economics, such as hypothecation of traditional medicines based on indigenous plant resources, get a heightened and somewhat emotional attention.

With reference to these parallel knowledge systems, there is growing concern that dominance of a single view of the natural world as expounded by modern science will undermine these civilizational knowledge systems. There is a fear that the process of globalization would threaten the appropriation of elements of this collective knowledge of societies into proprietary knowledge for the commercial profit of a few. Hence, an urgent action is needed to protect these fragile knowledge systems through national policies and international legislation, while providing its development and proper use for the benefit of its holders.

The dilemma of the developing world

There was a time when knowledge sharing was easy and crediting the original source of knowledge posed no problems. One can illustrate this by a number of cross-cultural examples. When the Arabs imported tamarind from India, they named it Indian date, because like the date they knew so well the new fruit also had a stone. To Europe, the numerals were Arabic, not because the Arabs had invented them but because the Europeans learnt them from the Arab world. For Europe, William Champion was the inventor of the process to produce metallic zinc, because it was he, who introduced it successfully. It did not matter that the process had long been known in China and probably still earlier in India; but this did not cause major conflicts.

India always looked at knowledge and especially useable knowledge in a dispassionate way. It is best exemplified by a controversy that arose in 1998 concerning Sir J.C. Bose, who is one of the founding fathers of radio physics. A story in the *Telegraph* carried the rather sensational title 'Bose invented Marconi's wireless'. The story was based on a report that the detecting device, called coherer, an instrument invented by Bose two years earlier was used by Marconi in his development of the wireless. One must understand the background.

As early as 1895, J.C. Bose demonstrated to an excited Calcutta audience the wireless transmission of radio waves over a distance of 75 feet through masonry. An early admirer of the Bose coherer was the British navy which used it to establish effective radio links between a torpedo boat and friendly ships. In May 1901 Bose wrote to his friend Rabindranath Tagore:

> The proprietor of a reputed telegraph company came himself with a patent form in hand. He proposed to take half of the profit and finance the business in the bargain. This multi-millionaire came to me abegging.My friend, I wish you could see that terrible attachment for gain in this country, that all engaging lucre, that lust for money and more money. Once caught in that trap there would have been no way out for me.

Exasperated by Bose's approach towards money, two of his lady friends, British-born Margaret Nobel (better known as Sister Nivedita) and American-born Mrs. Sara Bull on their own initiative obtained in 1904 an American patent in Bose's name. The irony of the situation is interesting. Here, in Nivedita one has a spiritualist advocating the cause of patents and royalties and a physics professor dismissing the idea. The reason must be sought in their backgrounds: Nivedita was a product of industrial Europe, while Bose was a child of the orientalized East. To take this point forward, it is interesting to note that while Bose had this attitude, the contemporary scientists in the Western world had a clear mind especially on usable knowledge, and that did not exclude the great Einstein. Many people do not know that with Leo Szilard, Einstein obtained about 40 patents on refrigeration. In a light-hearted way, I might emphasis that neither the refrigerator nor Einstein benefited from this exercise!

The context decides the content. There is a new context now, and the wealth making prowess of knowledge is being recognized now in the new India. Why do I say this?

Recently the list of 100 top billionaires of Indian industry based on net worth was published. The top five did not include the familiar names, except one. All the other four were people, who owned industries, such as Infotech or R&D intensive pharma. A new phenomenon is also beginning to be seen in India. An R&D savvy pharma company has developed a new molecule. It is undergoing experimental trials. The news that the molecule has moved up the innovation path from phase-I to phase-II enhanced the share value of this company, even though the actual molecule may not enter the market for the next few years. What does this phenomenon mean? This is the emergence of Indian knowledge industry.

The world's major growth industries—such as microelectronics, pharma, biotechnology, designer-made materials, and telecommunications—are already brainpower industries. These knowledge industries stimulate other industries, in turn, to become knowledge based.

Tomorrow's world will be dominated by knowledge industries. The emphasis in such knowledge industries is not on physical or tangible assets, but on intangible knowledge assets. Increasingly the traditional factors of production—land, labour and capital—have become less important when compared with technology; the economists have termed this as the 'expansion of the production frontier'. The value of intellectual capital of an industry will determine its rank and competitiveness. The nature of intangible assets will include several commonalities such as research and development, patents, proprietary technologies, databases, brands and even relationships, people and so on. In knowledge industries, there will be a major shift from people, who handled information and did routine and unthinking work, to those who will use knowledge at every stage. For such people, information and knowledge will be both the raw material of their labour as well as its product.

Why are the physical assets such as machinery in a factory, becoming less important? Plant and machinery are tradable commodities today. Even capital was a scarce commodity until recently and was used as a competitive advantage. But with globalized markets the companies around the world have access to finances at inexpensive rates. So even capital is no longer a scarce commodity. It is the intangible assets which are knowledge based, and that are non-replicable, unique and proprietary, that will provide the companies with a competitive edge.

For a cash starved country but rich in intellectual capital, like India, the emergence of knowledge industry is good news. But harnessing the full potential of knowledge industry requires an aggressive and visionary policy framework, creative planning, daring and risk taking. In particular, heavy emphasis on a stronger IP protection will have to be given to reap the full benefit of the emergence of knowledge industries.

TRIPs and technology transfer: the way ahead

Transfer of technology at fair and most favourable terms to the developing world has been highlighted in all discussions and debates on sustainable development. The Rio Declaration of 1992 as well as most of the multilateral environmental agreements emphasize the need for such technology transfers. The preamble of the World Trade Organization (WTO) Agreement affirms the objective of sustainable development in a manner consistent with the respective needs and concerns of members at different levels of development. Thus, an obligation is cast upon the WTO to bring about easy access to and wide dissemination of technology relevant for sustainable development. It is interesting to reflect upon what happened in the recently held Seattle round, and especially, what the US President said while addressing the ministers attending the WTO meeting in Seattle,

> We also must help these countries avert the health and pollution costs of the industrial age. We have to help them use clean technologies that improve the economy, the environment, and health care at the same time. And I will just give one example. Today is 'World AIDS day'.

And today our trade representative, and the Department of Health and Human Services are announcing that they are committed to working together to make sure that our intellectual property policy is flexible enough to respond to legitimate public health crises.

But the real issues, if I may say so, are truly complex.

Article 7 of the Trade Related Intellectual Property Rights (TRIPs) Agreement states, 'The protection and enforcement of intellectual property rights should contribute to the promotion of technological innovation and to the transfer and dissemination of technology to the mutual advantage of producers and users of technological knowledge and in a manner conducive to social and economic welfare, and to a balance of rights and obligations.' Furthermore, Article 8.2 states, 'Appropriate measures, provided they are consistent with the provisions of the Agreement, may be needed to prevent the abuse of intellectual property rights by right holders or the resort to practices which unreasonably restrain trade or adversely affect the international transfer of technology'.

Facilitating the access of developing countries to technologies required by them constitutes one of the key elements in accelerating the pace of their economic and social development. Such access is generally the result of licenses and technology transfer agreements. The fact of the matter is that the prospective technology seekers in developing countries face serious difficulties in their commercial dealings with technology holders in the developed countries. These difficulties arise for a variety of reasons. Some arise from the imperfections of the market for technology. Some are attributed to the relative lack of experience and skill of enterprises and institutions in developing countries in concluding adequate legal arrangements for the acquisition of technology. Some arisedue to government practices, both legislative and administrative, in both developed and developing countries, which influence the implementation of national policies and procedures designed to encourage the flow of technology to, and its acquisition by, developing countries.

Some of these difficulties may be overcome by suitable safeguards in the domestic IPR laws of developing countries. In addition, the transfer and dissemination needs of the developing countries have to be seen from the point of view of the capacity of those in need of accessing technologies, particularly where the cost of technology may be prohibitive due to economies of scale and other reasons. In such cases, in order to implement the related provisions of the TRIPs Agreement, commercially viable mechanisms need to be found. The denial of dual-use technologies, even on a commercial basis, to developing countries is another aspect that leads to the widening of the technology gap between developed and developing countries. Under this guise a variety of technologies and products are being denied, which could otherwise have helped to accelerate their growth process. This issue also needs to be carefully examined and seriously dealt with as a trade distorting and restrictive measure.

Intellectual property rights, biodiversity, TRIPs, ethics and equity

Harmonization of the intellectual property rights on the one hand with the TRIPs Agreement on the other hand with the implementation of the Biodiversity Conservation Act, are going to be rather crucial in times to come. Both raise serious questions of ethics and equity born out of inherent tensions between dominant concerns and value systems of the societies of the North and those of the South. In the perception of the south, the principal concern of TRIPs is the strengthening of patentee's rights in the global market. This arises from the TRIPs requirements to extend the scope of patentability to all inventions, whether of products or processes, in all fields of technology. How some of these will serve the public interest, particularly in vital areas like availability and price of medicines and agro-products and protecting the domestic industry from unequal competition is a matter of concern that many developing countries worry about; India being no exception.

The issue of biodiversity is even more complex, contentious and tends to get emotional. It is inextricably linked with the means of daily survival (food, health, shelter, energy, etc.) of about 70 per cent of India's population and thus to the rights of the communities. Here one has a clash between two very different value systems. The traditional societies in the developing world value diversity of life and plurality of knowledge systems besides plant-animal symbiosis and a harmonious adjustment to the environment as the cherished way of living. The information as contained in nature is their principal property and they view the unlimited community as the rightful beneficiary of human knowledge and endeavour. In contrast, a modern, developed society in the typically Western mould recognizes only one valid knowledge system, based on rigorous demonstration of causation and a common framework of communication. The nature of knowledge is intellectual and it is recognized as a private property. The beneficiary is one who can pay according to market forces. Diversity is no more than a raw material. The patentee has monopolistic rights, and though there is a sea change in the items sought to be patented, the criteria of patentability have not changed for over hundred years. The concerns of the indigenous populations concerning the loss of both their material and knowledge, as biodiversity is increasingly appropriated by a few and natural resource stocks deplete and disappear, these concerns will have to be addressed urgently and seriously.

Recognizing the impossibility of preventing cross border flows of genetic material in the age of globalization, one school of thought considers it prudent to build strong data bases on biodiversity and use them to levy charges on bioprospecting and royalties for their commercial use for sharing the benefits. This stand of opinion would welcome inflow of science and technology inputs from outside to add value to resources and export for global market share. The counter viewpoint sees this reasoning as succumbing to a colonialist option under pressure and argues that in the absence of legal rights for traditional innovators and the collective right of the communities as owners of biodiversity, bioprospecting degenerates into 'biopiracy'. The truth surely lies in between. In order to change the notion of 'biopiracy' to the notion of 'biopartnership' between the North and the South, one needs to move towards TRIPs plus which implies TRIPs plus ethics and equity, the latter taking care of the concerns mentioned earlier.

A fresh look at national innovation policy

With increasing globalization, privatization and corporatization, a question that is engaging the attention with regard to the process of innovation is, where do the capabilities for innovation reside. Till the other day, it was believed that they reside in nation states. Certain nation states were able to create the climate and provide the infrastructure which spurred the innovative genius; certain other states failed to do so. This led countries to devote thought to building up a national innovation system and evolve a national innovation policy going beyond the national science policy or national technology policy. A national innovation system was seen to be embedded in larger institutions, labour market, government policies and biases, regulatory framework, legal system and the way these elements related to and interacted with each other. While the idea of the innovation system has not become invalid, the national part of it is losing relevance as R&D gets increasingly internationalized, the world becomes more and more integrated and the idea of the autonomous nation state is fading.

Different nations regard innovative capabilities as residing essentially in firms as these are characteristically a function of the inter-twining of scientific and technological capabilities of a firm with its organizational and management capabilities and strategies. Some even thought that innovation flourishes in certain geographical regions with special attributes of history or talent, like Silicon valley in the USA, or Bangalore in India, while some view global networks as the essential seat of innovative capabilities in the emerging R&D paradigm. These rival views are not necessarily mutually contradictory and they may be valid within their limited contexts. However, they do not point to the fact that while it may be appropriate to think in terms of 'technology systems' that encourage or inhibit a culture of innovation, it may be a bit artificial to seek such systems within the political boundaries of a nation. I will deal with this important point a little later again.

Innovation in a global economy

Advances in communication and information technology have connected the people and the places of the world as never before. It made the vision of a global village a physical possibility. This was aided by the end of a bipolar world following the collapse of the Soviet Union, and the aggressive pursuit of its economic interests and ideology by the developed world. The reality of globalization is very evident in the sphere of economy. International economic integration has proceeded apace as seen by the galloping world trade, integration of financial markets and the changing character, increasing clout and expanding activities of the multinational corporations (MNCs).

An interesting feature of world trade is the emergence of three powerful economic groupings, viz., the USA led NAFTA, the European Union and the Far Eastern group of countries led by Japan. All the three groups are technological power houses. That leaves out most of the developing and under-developed countries which have to evolve their own strategies to survive in a ruthlessly competitive world. The most far reaching changes for global integration are, however, being wrought in the area of international finance. It has made 24 hour global trading a reality; it has also made it easy for foreign institutional investors to follow market developments and respond. It has facilitated internationalization of portfolios and helped the banks and securities houses in their role as managers of large scale capital flows. The international finance sector has been extremely innovative creating entirely new financial products like Swaps, Futures, and Options for the market. The economical globalization obviously needs effective regulatory structures for global finance, but such political structures at the international level have not emerged yet.

One would do well to take note of the geographic dispersal of R&D activities by multinational corporations (MNCs). It is important to understand the causes, nature and implications of this trend because technological change is central to economic growth. Also it is necessary to comprehend the corresponding structural changes that would be necessary to capitalize on the emerging opportunities. A fundamental change seems to be taking place in the nature of the MNCs. They are no more tied down to a home country in terms of a base for organization, capital and R&D. Increasingly, they are ceasing to be the controlling and coordinating centres for a set of peripheral and independent national subsidiaries and acquiring the character of a network where national units are viewed as sources of ideas, skills, capability and knowledge to be harnessed for the total food of the company. International exploitation of national technological capabilities by major firms is on the rise. I shall discuss this issue further.

Research as a global business

The compulsions of international trade and commerce and the phenomenal advances in information and communication technology have made a globally integrated world a reality and there can be no sequestered islands in today's world. As technology is the prime mover for a modern economy, the economic integration of the world cannot be complete if industrial research is not globally integrated. The globalization of R&D implies that the premium will be on research as a means ofwealth creation, pursued as a business proposition, seeking exploitable knowledge in worldwide collaboration.

The globalization strategy now includes taking over of R&D intensive corporations. The Japanese, for instance, have promoted a new concept of techno-globalism, which is being interpreted as 'the strong interaction between the internationalization of technology and the globalization of the economy'. It actually implies 'widening cross-border interdependence between individual technology-based firms as well as economic sectors, especially through the restructuring of high and medium-tech industry'. Incidentally, it is interesting to note that the intensity of such transnational cooperations and setting up of R&D facilities is the highest in high-tech areas, such as electronics, biotechnology, automotive industries, etc.

The innovation chain will increasingly cross transnational boundaries today. Companies realize that to gain competitive advantage, they have to leverage their capabilities. Many companies across the world today consider it to be rather unwise to strive for self-sufficiency in technology development, particularly in an era, where the R&D costs are increasing rapidly. With trade barriers among the countries disappearing fast, companies have to provide the best technology globally to their customers. As a part of the global innovation strategy, several companies world over are scouting for new ideas and patents, which the originator is unable to exploit for a variety of reasons.

There is another interesting and subtle aspect to the whole game. The globalization of R&D is closely linked to globalization of business and consequently to global competition of skills. The competitive advantage in high-technology business increasingly depends on the underlying technical skills of the business rather than on particular products. As product life cycles are becoming shorter, skill life cycles become longer.

The product then is merely an intermediary between the company's skills and the market it serves. Rather than being the focus of corporate activity, products are actually transient mechanisms by which the market derives value from a company's skill-base and the company derives value from the market. The high technology companies are therefore asking as to what skills, capabilities and technologies should they build up, rather than asking a stereotype question, as to which markets should they enter, and with which products.

Several factors are helping to accelerate the globalization of industrial R&D, but the most important factor is the creation of 'seamless laboratories' around the world, is the evolution of global information networks.

In this emerging scenario, India is poised well to become the global R&D platform of the twenty-first century. Why is this so? First and foremost: The costs of doing research are lower in India. It is remarkable that the entire science and technology budget of India, which includes space, defence, atomic energy research, does not exceed 2.5 billion US dollars, whereas the budget of Siemens alone is 5 billion US dollars! This cost-effectiveness gives India a strong competitive position. Secondly, the high quality science base prevalent in India in certain select areas is a big attraction, especially when one recognizes that industrial R&D is be-coming increasingly science based. Thirdly, the companies perceive that such collaborations provide specific mutants of an existing technology or a new technology that is fully adaptable to the distinctive conditions prevailing in India. The fourth factor is the perception of the new environment in the post liberalization era in which Indian R&D institutes and industrial firms will operate, where factors such as improved freedom and flexibility, better communication and IPR protection, improved consciousness about quality and time of delivery, will improve dramatically. Paradoxically, the prospects for North-South partnerships in research and technology are thus looking more attractive than ever before.

The politics of a global village

With the integration of the global economy, the world is on its way to integration in other spheres. Some features of the post bipolar world are already distinguishable.

There are enough pointers to indicate that only certain political philosophies and structures are acceptable to the global community. Human rights are understood in the Western context, and a democratic polity are the prescribed dispensations. Concerns of national, regional and global security are coming to the fore and new alliances are in the making. The threat of an all out nuclear war seems to have receded, though there is a proliferation of nations having possession of or access to nuclear power. The number of regional conflicts has expanded and though the super power arms race has stopped, the motivation of smaller powers to acquire military might through acquisition, or indigenously developing weapons and systems of modern warfare has increased. The UN is on the decline as an overarching global mediator.

A net result of the strivings for globalization is the weakening of the nation state and a dilution of the concept of national sovereignty. States are becoming increasingly less important as political units to control whatever phenomena—economic, sociological, environmental or technological—taking place in the world. The erosion of national sovereignty is seen in the proliferation of regional trade and economic agreements, growing economic power of the MNCs and the conditionalities imposed by international financial institutions for accessing resources under their control. These developments are redefining issues of power and authority over the control and management of resources for development and bringing to the fore inherent ideological and cultural tensions.

Information technology and democracy

One is in the early stages of a digital revolution that is changing the way information is generated, owned and used. This revolution will enable democracy to be more integral to all human endeavours, one's government, business and personal lives.

The basis for democracy, and a prerequisite for its long-term success, is an informed public. The fundamental political changes in Eastern Europe, Ireland, East Timor, and South Africa were all possible because, try as they might, those governments were unable to control the generation and distribution of economic, political and social information within their societies. Once that control is lost, democracy is inevitable. Chinese students found that fax machines were more powerful than rifles.

First, the people must be literate enough to be able to read and understand the information that is relevant to their lives. Television and radio have been revolutionary tools in this sense. But the written word, plastered on a wall, passed out on the street corner, remains an essential tool for social change. People must be able to read and write in order for democracy to succeed. One needs to take note of this in India.

Secondly, democracy requires that people have open access to certain basic economic and political data and to a diversity of opinions. This requirement means access to information for all.

People must be able to share their ideas with others. Information technology is making it possible to reach beyond time and place to a wider community of citizens. The content creation is, therefore, going to be a crucial part of the strategy, since information could now be used for a variety of purposes, from propaganda to enlightenment.

Despite the crucial importance of access to basic social and political information in the democratic process, there is no public agreement concerning what information should be broadly accessible. May be one needs to develop a Bill of Rights on Information, which will establish the public policy concerning the accessibility of certain basic information to all citizens, dealing with both their rights of access and rights to privacy. The recent movement started by Anna Hazare in Maharashtra is interesting from this view point.

Proponents of the current intellectual property system maintain that only the market economy can provide the appropriate incentives for efficient and effective distribution of information. They argue that large international publishers have contributed to making information a global commodity; that without the investment of these companies, there would be less information, it would be of lower quality, and would cost even more than it does, and that which did exist would be less accessible to potential users.

Not all information, however, is adequately served by the market economy. The basic infrastructure of knowledge, including history, philosophy and the arts, as well as minority opinions are essential to a healthy democratic society. These areas require a 'social investment', on the part of the community.

A monitoring system to measure the quantity, quality, and impact of different kinds of information upon the economy and society is needed. One has such information with regard to literacy today but not with regard to the issues of access and interaction. One needs a set of indicators to serve as a basis for regular monitoring and reporting on the state of the information systems.

Technology is not value-neutral. Certain technical developments are more likely to serve an open democratic process than others. Governments should create and implement a development strategy for information technologies that is supportive of democratic processes. Such a system should explicitly address the three goals of literacy, access and interaction, and the roles of the private and public sectors in promoting those processes.

Finally, it is important to note that the advance of information technology will not only force a review of the legal and practical meaning of the notion of individual privacy, but also that of the 'privacy' of nations. Furthermore, the existence of inexpensive, multiple and worldwide networks for communicating information will shift power from government to individuals. The netizens will have a world of their own. While information technology will vastly increase the power of a government to monitor its people, the government's control over the information distribution will be diminished. Information will still be power—but it will be a shared power. The increased worldwide distribution of information will also lead to increased awareness in less developed countries of how the people with superior life styles live. This will provide interesting driving forces, both nationally and globally. The availability of information networks will diminish the reliance on elected representatives and technical experts to make decisions on behalf of the public at large. It will be the forum of public opinion rather than ministerial negotiation that will control the final decision. This will also mean that greater attention to 'quality' of information available to the electorate will be important.

Finally

As one comes to the end of this millennium, it has become mandatory for every speaker to speak about the new millennium scenario. With the rate of knowledge doubling up every five years, it is becoming difficult to predict the future. As someone has said, 'A technology of the twentieth century' symposium held in 1895, based on the level of knowledge that existed at the time, might not have mentioned aeroplanes, radio, antibiotics, nuclear energy, electronics, computers or space exploration! What are the equivalents of this that will be missing when one makes predictions in technology, for the next century?

In the first half of the twentieth century, physics occupied a dominant position. But biosciences displaced physics in that leading role now. It is not cellular and molecular biology alone, but also ecology, that is taking biology to a different level of impact. The technological potential of biology will pose a new challenge due to its intimate connection with agriculture and health, areas of great political sensitivity and social importance. Technology transfer in biological fields does not relate to just sharing the luxuries of a consumer society but also to sharing the means of survival. The issues in trade and technology negotiations thus become far more complicated. The new shifts in the coming century will be hard to predict, but it is quite clear that the continuing wars of the information revolution and the felt impact of the 'gene revolution' in shaping the future of mankind will have a profound impact on the global economic and sociopolitical scenario, that will be hard to anticipate today in its full measure.

Although I cannot predict the new millennium, I can certainly express a wish for a new millennium. Let me hope that no matter what progress science and technology makes, one will be able to network not only the minds but also the hearts of the 'global citizens'. That would certainly mean a great future for the globe in the new millennium. If Bose and Einstein, those two great 'global citizens', were alive today, I am sure, this is exactly what they would have wished.

Content Counts: Science Policy for the Twenty-first Century

UWE THOMAS

When you prepare a lecture like this, you read up a little bit, in order to become more familiar with the two godfathers of this series and in particular with S.N. Bose and his scientific career. In fact I was surprised to learn that in the year 1917 S.N. Bose and C.V. Raman both taught at the same University College of Science, Calcutta. It must have been a great time for science. Even though, when Bose sent a letter to Einstein with his article on 'Planck's Law and the Hypothesis of Light Quanta' for the German scientist it must have seemed like coming from nowhere. Einstein, one of the group of physicists who revolutionized physics by discussing radical theories, was not familiar with Dhaka University in India where Bose was based, but he read the article and recognized its value. He translated it, sent it to the *Zeitschrift für Physik* and then he sent a postcard to Bose (see pp. 26, 27) to tell him that his article was to be published soon and that he liked it. Now can you imagine, there was no Internet at that time, a postcard was sufficient. So what does one learn from this story? One learns a very simple fact that it is the content that counts, not the mode of transmission.

When we talk about science and culture—both of which are interrelated—we are still astonished when we learn what was happening at the beginning of this century. The questions these physicists were asking were really unheard of, and when we read about that period, we are surprised how far they went in

*Tenth Bose-Einstein Lecture, 26 October 1999, India International Centre, New Delhi and 30 October 1999 at the IIT Madras.

their ideas and how radical they were. The things that counted in science were curiosity and radical questions, and they are just as important today as they were a hundred years ago. The value of basic research is therefore a cultural value, and if research policy has a duty, then it is to create opportunities to ask these radical questions. There is a lot of talk about applied research today, and it is even true that what used to be compartmentalized into basic and applied research is now moving closer together. There are mathematical questions which used to be regarded as basic research, but are now really applied and crucial for industry. But still, the value of basic research, the possibility of asking radical questions, determines whether or not a scientific culture exists.

II

In this world full of conflicts, there is sometimes a feeling that preservation of cultural values and fundamentalist attitudes are close to each other, but the contrary is true. They are the opposite of each other. This is an important realization when one hears and reads about the clash of civilizations. Should more people believe in that theory, this world may not have a real chance of survival.

When I came to India, some lost memories returned. I remembered as a boy hearing about Mahatma Gandhi and the possibility that one can fight a world power with one's soul. The admiration young people had at that time for this man came to my mind. And I thought that I might have a look at the book, the *Bhagavat Gita*, I shall quote from it, namely what Arjuna said when he was talking to Krishna before fighting a battle. He said, 'Better for me if the sons of Dhritrashtra, weapons in hand, were to kill me unarmed and unresisting on the battlefield'. Now, for a general this is unheard of, and maybe he did not sustain that conviction. But this is what one admires in religions, both in the Christian and the Hindu religions, that they promote values which are difficult to follow—and one does not follow them all the time—but which gives one a direction far removed from the fundamentalist ideas of the present time.

Let me narrate my experience of a competition of theatre groups from German schools held in Berlin. About 300 schools participated, 10 of the best ones were selected, and I had the pleasure of opening the final performance. There was a group of young students from a school in Hamburg whose play was called 'Nathan der Weise'. Most people know what this play by Lessing is about; it is about tolerance. But they did not perform the play; they did perform something about it. This group of young people included a girl from Croatia, a boy from Serbia, a young Russian—there were all kinds of nationalities, as is now usual in German schools. And they acted out their own lives, in a way, and discovered tolerance. The teacher who worked with them said to me, 'We were working for two years, but it was our chance to bring the young people together.' When one remembers what was happening at that time in Yugoslavia, one can imagine what that meant. What I think is the main idea of science as part of culture in the present time is that it teaches tolerance and respect. These are the two most important virtues of the twenty-first century.

III

What is the challenge of the twenty-first century? The six billionth human being on earth was recently born in Yugoslavia, where a baby was thus selected. A United Nations study estimates that by the year 2050, there will be 9.4 billion people living on earth. Let me quote Paul Claudel, the diplomat and catholic philosopher who said, 'Before we change the world, it might be more important not to destroy it.' Maybe he is wrong. Maybe it would be better to say, 'We have to change the world in order not to destroy it.' The consumption of crude oil per capita in the United States is 7.9 tons per year. In India it is 0.26 tons. The consumption of paper per capita in Germany in the information age is 192 kg per annum, while in China it is 27 kg. Now let us imagine for a moment that India or China demanded the same proportion of resources as the United States or Germany. Obviously, with regard to paper, one would have to forget the forests, and with regard to oil one would only have enough resources to last a few years. There are some politicians around the world who understand the problem. A famous politician who aspires to become the president of the United States, Al Gore, has written a book called *A Marshall Plan for the Earth*, in this book one will find everything necessary for this planet to survive. Unfortunately, his own people don't follow his advice. He himself has not implemented what he demands. Of course, he believes in equal rights to resources for everybody on earth, but the belief alone is not sufficient—it has to be implemented. The problem faced by most politicians is that even if they have good ideas, if people do not follow them, they have to either choose to change their minds or to forget being a politician, which is unfortunate. What one needs to do, if one wants to implement this modern version of equal rights to resources, is to convince the people and act accordingly. The need of the hour is saving on conventional resources and the development of alternative energies and raw materials.

IV

My fourth point relates to the potential for successfully meeting this challenge, which of course, involves education. I was very impressed by the statement of the Minister of Human Resource Development, Professor M.M. Joshi, about reducing the rate of illiteracy in India from 43 per cent to 25 per cent. This is a major task to be completed in five years. This is as important for India as it was for mankind to go to the moon. We should admire India for being one of the few democracies in the world which is functioning with such a high rate of illiteracy. Participation in elections in India is higher than in the United States, an old democracy which also experiences some trouble with illiteracy, though the percentage may only be 10–15 per cent. Responsible acceptance of democracy is the basis for education which in turn is the basis for democracy—they are interlinked and India is a very interesting example. This is one of the reasons why I believe this country has a great future. There are other reasons. When I see the high level of science in India with German students in attendence at the Indian Institute of Technology in Delhi, headed by Professor V.S. Raju, I realize that this high level of science studied by brilliant people shows great promise for the future. I believe that this will help India to become one of the leading economies within a generation.

A few weeks ago the German Government decided, after long and elaborate discussions with industry and science, on an action programme for development in the information society, which can be found on the Internet on the BMBF's web pages (http://www.bmbf.de). It is a comprehensive action programme covering all areas from regulation via education to research in that field. One part of this programme is, I believe, worth mentioning, it relates to the way education is changing. Education for centuries was a process of lecturing to people followed by discussion. The main stay of education was the printing technology, which is now rapidly being taken over by electronic technology. The education process itself is undergoing a profound change, and it is quite interesting to note how this change is viewed in the United States. Education is looked at as a market, a global market to be precise. It is said that it has all the qualities of a growth market (about the same size as health care, which triggered many changes in biotechnology, a former growth market). It is characterized by dissatisfied consumers who are not happy with the education provided, so far a relatively low utilization of technology, but with a great potential for the application of sophisticated technology, and finally strategic importance. In discussions about this issue it is interesting to learn that this market is worth about 9 per cent of world GDP or 3,000 billion dollars worldwide, an enormous figure. One of the main markets aimed at is Asia.

One may recall Michael Millken who used the concept of junk bonds and made a fortune. Unfortunately he came into conflict with the SEC, which supervises the stock market, and as a result he had to go to jail. He was in prison for some time, but he was able to keep most of his money. The reason why I talk about Michael Millken is that a few months ago he founded a trust, together with a few other people in the United States—a company called Knowledge Universal Incorporated. They put one billion US dollars into it with the belief that the hottest area for venture capital over the next decade will be education. The British are also thinking along these lines—a leading British politican in an article said he wanted Great Britain to become a world-beater in this new global industry. I liked that word—a world-beater.

Of course every country is thinking about this issue. If you read the action plan I and II you will find that in Germany we believe that the Internet will become an educational tool of strategic importance for all who offer education worldwide, but one also knows that this will only work if one defines the changed role of teachers, and in particular if one trains them to make use of these new possibilities. Secondly, this will only succeed (as I said at the beginning, the content counts) if the contentware, which is not software in the classical sense but something quite different, is properly designed and its quality guaranteed. It is true that in this field we shall face global competition and there is no doubt about that. The government is just in the process of starting a programme which will cost it about 400 million DM over the next four years to support the development of such contentware. It may take 10–15 years for dealing with the variety of subjects, but at the end it will have profoundly changed the educational process. One will be able to learn partial equations with the help of computer contentware, but this will only be possible if one understands the learning process much better. I don't know about India in this field, but I must admit that in Germany, where the universities train a lot of teachers, research on the learning process is a scarce resource. It is like having a big company and putting less than 0.1 per cent or maybe 0.01 per cent of the turnover into research. One does not understand the research process very well. It is a complicated issue, as it probably varies in different fields, and there has to be specialization, but carrying out research into the learning process will be one of the activities necessary to make full use of these new possibilities.

There is still another dimension: When one reduces education to a market, one misses this dimension. In the recent Systems Exhibition in Munich an American company called the Learning Company, which is the market leader in the United States, offered computer-based learning aids, all Walt Disney! Now, I have nothing against Walt Disney, I like Donald Duck in particular and his role. But I believe that cultural values and education have something to do with each other. And I think a world which is not uniform but varied is much more interesting. I therefore hope that this question of the education market will be challenged by those who believe that education is part of the culture of a country. This does not mean that one is not competing in a market, but that one must remember one's responsibilities in preserving cultural identity in a fast changing world.

V

My fifth point concerns the general question of the global market economy and regional cultures. Not long ago, there was a very heated discussion in Germany on the question, whether teaching at universities should be done in English in some fields. Now some courses are offered in English at German universities, and I have nothing against this. Germans are very lucky in Europe, at least in continental Europe, because they are forced to speak more than one language, and that is an advantage rather than a disadvantage for them. When one goes to Britain or the United States and meets people who speak only English, it is a pity, for they miss out on something in their lives. But it doesn't mean that one should not use English as a global language, as it now serves in scientific communication.

But related to the subject of the global market economy and regional cultures, is another problem which was addressed very clearly when I visited the state of Israel recently. This is the growing gap between rich and poor within a country. I visited a total of four universities in Israel, and was quite surprised that students and teachers at all these universities were deeply concerned about the coherence of their country and regional Israeli culture that would be lost because of this growing gap between citizens.

I don't know whether one who has been to Stanford University ever tried to see Palo Alto. Palo Alto is a beautiful and wonderful city. Some years ago I had rented a car there and unfortunately I lost my way, crossed a railway track and ended up in the other part of that city. When I returned to my hotel, people said, 'You went there? Do you know that this part of Palo Alto has the highest crime-rate in the United States?' Palo Alto is beautiful and Stanford University is wonderful. But to have these two completely separate societies side by side appeared to me as a terrible challenge destroying the notion of regional culture and identity.

I don't want to be misunderstood. I am a fan of the United States. I love to be there. I like the people, their friendliness and the way they are not as complicated as the Germans are. It is true, when I visited the Massachussets Institute of Technology a few years ago, I found an old MIT yearbook dated 1900, with an article by a certain Professor Swain about his visit to Germany. In fact he was quite unhappy about the Germans. He said that the Germans invent a problem and then try to solve it by elaborate mathematical analysis. In some ways, his observation still sounds familiar to me. Germans are strange in one way or another, aren't they?

Anyway, I think there is a deep conviction in Germany which

cuts across party lines and is a part of the national consensus namely that companies have a responsibility for what has been called the stakeholder value. The chairman of Daimler-Benz (not yet DaimlerChrysler at that time) once wanted to go up in the stock market, and he said, 'For me only the shareholder value counts'. He repeated that exactly for two months, and then he turned around and in a great speech he said, 'Stakeholder value is more important'.

We Germans have this common denominator: We reject the notion of two societies in Germany, of those who have and those who have not. We do not want to allow an utterly poor section of society, separate and criminalized because people have no choice, and are perhaps relatively successful in keeping the equilibrium, as we believe in solidarity between people.

This does not however mean that the idea of stakeholder value should prevent Germans from accelerating structural change. On the contrary, it is believed that we have to repeat a process that occurred in the nineteenth century in Germany. At the beginning of the second half of the nineteenth century, there was one highly industrialized country, particularly in the textile industry—that was Great Britain. They were the leaders. The whole process of producing steel, I must admit, was stolen by German entrepreneurs. Germans like Hoesch and Krupp went to England, looked at the technology, wrote it down and then copied it. This is hopefully not something that Germans are going to repeat. At the same time however, thanks to a far-sighted policy technical universities were introduced, and—it is difficult to imagine today what this meant— technical universities could award doctoral degrees. The old universities like Göttingen and Heidelberg protested and said that the mission of technical universities had nothing to do with *Wissenschaft* as science is called in Germany. *Wissenschaft* in Germany is not only natural science but includes humanities. The technical universities and research facilities which were established at that time were just one factor towards progress, another one was deregulation.

In the year 1870 the deregulation of public corporations and the banking business took place in Germany, and all three large banks of the present time—Dresdner Bank, Deutsche Bank and Commerzbank—were established within two months in this period. Most of the large companies responsible for the wealth in Germany today were founded at this time. AEG-Telefunken grew within ten years from zero to sixteen thousand employees—what a leap! There was a stock market which financed that kind of expansion providing the capital, and as a result, by the end of the nineteenth century Germany had all the new industries. At that time 50 per cent of all electro-technical production in the world and 80 per cent of artificial colours came from Germany. This was the result of research, venture capital and the stock market. Of course, it was also the result of entrepreneurship, but it is significant that many of those entrepreneurs like for example Robert Bosch had a sense of social responsibility. In some parts of Germany one can still see the colonies with small houses, quite nice, with gardens and so on, which were built by these entrepreneurs for their workers because they felt that they had a social responsibility.

Now, what can we learn from this today? Of course, we have a completely different situation, but a recent phenomenon, not very well known, seems to me quite similar. It is not very well known because it is a rather new development: we now have a very innovative stock market in Germany, called the New Market. The NASDAQ, set up 21 years ago was a very important instrument for the provision of capital for high-growth companies. Almost all the famous high-growth companies in the United States got their capital through NASDAQ; without it there would perhaps have been far less high-growth companies in the United States. Now, not long ago, a stock market of this type has been set up in Germany. It happened in March 1997. Surprisingly in 1999 some 150 new high-growth companies went to the New Market to get their capital, to make an initial public offering, as the Americans say. This compares to the NASDAQ figure of between 350 and 500 per year. So per capita the New Market in Frankfurt, which is part of the Deutsche Börse AG, compares already to NASDAQ within one and a half years. Whether this market keeps up its dynamics remains to be seen, but it looks like it will since there is high liquidity to be found here.

I am talking about this because new companies are needed in Germany, new markets are best covered by new companies. Of course a strong research base is also needed, which I think the country has, as well as experienced venture capital, which developed in the last few years together with a stock market, which provides the capital for those who have a high growth capability, and finally one needs an entrepreneurial culture accepting this. A culture where young people are prepared to take a risk and set up their own companies. If one asks students at technical universities in Germany today what they want to do, one will not very often get the answer, 'I want to go to the research labs at Siemens', but 25 per cent of them will say, 'I want to set up my own company together with friends'. This is a cultural change, and Germans are very pleased about it, because they believe that technology transfer can only be done through people. It is people who can transfer technology from their research work into a company.

Throughout the 1990s there has been a lot of talk about the service economy. When one went to the United States at the beginning of the 1990s one was told: 'You Germans do not have enough people in the service economy, and therefore you will not be successful'. What is success in the global economy of the present time? By the use of trade figures, one sees that Germany appears quite successful, more than the United States. There is no economy which is purely a service economy. Economies which are successful in terms of trade figures must have a strong manufacturing base like Japan or Germany. It is increasingly true that one cannot separate the two. But a strong manufacturing base matters, if one wants to have balanced trade figures. It is a tragedy for the United States that it has lost part of its manufacturing base. This occurred at a time when President Reagan had driven the dollar much too high, so that manufacturing companies in the US had trouble to remain competitive. It is a tragedy because all are at risk due to the high deficit of the United States balance of trade. This could cause imbalances across the whole world if confidence were to be lost. So let one remain confident that the United States will regain long term stability and fight their trade deficit successfully.

VI

To conclude; of course, everybody knows that all countries share a global responsibility as we are one world and we have only one world, and the main problem is to convince the people of this truth. If we cannot convince our people about the risks this world is running, we may lose the game. If China and India would take the course of resource consumption as in the United States in the next twenty years, there is little doubt that the earth might be in danger since we have only one earth to share. And if this is true, it means we need a new policy dimension both in education and research, which leads us to a common understanding, to a new dimension of cooperation.

One may think of India, a big country with one billion people, assuming that just 10 per cent of them have comparable living standards to most Germans, then in a way India contains an embedded Germany. If one considers the potential of this big country, with so many human resources and its great tradition in science, and then looks at Germany, which is one of the largest producers of investment goods in the world, and therefore has to bear a global responsibility, it seems to me that both countries together could develop a common vision, to make the twenty-first century a problem-solving century on the basis of equal rights to resources. It is our hope and optimism that this will be possible, in order for the earth to survive.

German Research and Technology: Challenges and Perspectives in Times of European Integration and Globalization

KARSTEN BRENNER

Indo-German cooperation

The Madras IIT is indeed an appropriate place for a dialogue on the role research and technology, education and science can and should play for the political, economic, social and cultural development of India and Germany. This role is more important than ever in times like the present, marked by not only deep structural economic changes and widespread unemployment, but also by growing regional cooperation and worldwide interdependence.

Successful cooperation has to build on mutual understanding of the partner's specific situation, its basic position, concerns and objectives. For this reason it may be of interest to the reader to get an idea of the German policies on science, research and innovation. It is no surprise that many issues and challenges that confront Germany are very similar to those in India and that its approach to reforms is often the same.

Political and economic context

Germany's political and economic situation can be characterized by five major trends and challenges:

THE PROCESS OF GERMAN AND EUROPEAN UNIFICATION AFTER THE FALL OF THE WALL BETWEEN EAST AND WEST

This historic event is not only of fundamental importance for Germany, but its neighbours in Europe as well. With its

*Seventh Bose-Einstein Lecture, 30 April 1998 at the IIT Madras.

unification Eastern and Western Europe are growing together again. Germany is resolutely engaged in restoring contacts and building new bridges in order to reunite the continent. At the same time it is supporting its eastern European partner countries in their restructuring processes.

In the former communist countries, research and technology, and education and science played, a particularly important role for bringing about economic and social modernization. Eastern Germany went through a major reorganization, involving reform of higher education and the modernization of research institutions. While the state supported research capabilities have by now reached the same level as in the western part of the country, research and development activities in the industrial sector are significantly lagging behind, due to lower productivity levels. Therefore, R&D in private industry requires a major adaptation effort.

THE PROCESS OF EUROPEAN INTEGRATION

The European single market has been completed in 1992 and it is a firm German position that it must be kept open to the world to avoid the emergence of a 'fortress Europe'. At present there are three new themes on the agenda of the European Union:

— The introduction of a common currency, the Euro.
— The step by step enlargement of the Union to include Eastern European countries.
— The internal reform of the Union aimed at more efficient decision making processes, the reform of the agricultural policy, an overhaul of the system of finance and the Regional Fund, which gives structural support to less developed European regions.

Cooperation in research and technology, and higher education play an increasing role on the European level.

THE PROCESS OF GLOBALIZING AND WORLDWIDE ECONOMIC RESTRUCTURING

Globalization for us, means much more than global trade in goods and services. Global trade, of course, is nothing new, Germany and India had trade links indirectly for hundreds of years. The characteristics of today's globalization, helped by modern communication and transport technologies and the elimination of trade barriers, seem to be the following:

— To a growing extent companies are investing, producing and competing worldwide and their interaction is a mixture of targeted cooperation and tough competition.
— In their investment decisions the managements of companies compare the potential locations worldwide in terms of costs, available knowledge, qualifications and the market potential.
— Key company functions are no longer tied to central locations.

Research and development, planning and logistics, market and product development now take place at the most favourable locations, i.e. in high volume markets or wherever economic and technological competence and suitably qualified personnel are available.

This process has adversely affected the German economy, as well as its labour market. It gave rise to a public debate on the weaknesses and strengths of the economy and on the attractiveness of Germany as a location for production and investment. This discussion was all the more intense as, after an initial boom due to the stimulus of German unification, the economy was hit by recession which led to high unemployment, a phenomenon not experienced so far in post-War Germany—which will be only overcome now.

A general consensus has emerged that globalization not only poses a threat but also offers great opportunities. The prerequisite is, however, that we must respond actively to the new challenges, undertake long needed reforms, abolish over-regulation and over-bureaucratization, increase innovative capacity and make the labour market more flexible.

THE SHIFT FROM INDUSTRY TO SERVICES

The fourth trend is the shift from an industry-based society to a society marked by services, new information technologies, and a growing communication and media sector. The importance of the role of knowledge on which successful economic activity has to be based, is increasingly being realized. The movement towards a knowledge based society is giving globalization its real drive, a drive that makes innovation the most important lever and instrument of modern economic policy.

The challenge facing Germany in this respect is threefold:

— It can no longer rely only on its traditional strengths like mechanical and automotive engineering, and chemical products that are based on advanced technology. The country has to also achieve a stronger position in such goods and services based on so-called cutting-edge technologies and on generic technologies like microelectronics and biotechnology.
— We have to make more use of the innovation potential in the service sector, not only as user of new technology, but also as provider of production related services.
— Finally, we need to bring in structural change and innovation through launching a wave of new technology-oriented firms.

Given Germany's relatively high labour costs, its economic success and new employment opportunities are more and more dependant on research intensive products that succeed in the worldwide innovation competition. A recent editorial in the *New York Times* succinctly described the US current economic success as follows, 'We are leading in everything that is light:

microelectronics, software, pharmaceuticals, computers, aircraft'. In other words, success has been achieved through knowledge-intensive products.

SUSTAINABLE DEVELOPMENT

The fifth challenge to Germany, as to other countries, is to contribute actively to the sustainable development of the world. The issues raised in the Rio Agenda along with the ambitious domestic targets set, define an important part of the country's research programmes, policies and budgets, penetrating both on the national as well as on the European level. Stringent environmental standards are not a burden but have given rise to a new industry which offers highly sophisticated technology both for healing as well as preventing damage to the environment.

High priority for innovation and research and development (R&D)

Consecutive to these trends and challenges a new consensus is emerging in Germany on the importance of innovation as the key to solving current problems that will actively shape the future of the country. Innovation implies economic, social and cultural sense on all levels: government, industry, research institutions, and last but not least in the attitude of the individual citizen. Enabling innovation in this comprehensive sense has been the guiding principle of the past and the present federal governments as well as our Länder Governments. Research and technology, education and science have become the decisive factors for increasing productivity and economic growth. They have also become the only promising approaches to overcome unemployment and solve the environmental problems. Formulated in economic terms the third factor of growth: The production of knowledge and it's efficient and effective use through research, combined with high qualification of the labour force and efficient management become the key factors at times when capital for profitable use and skilled labour are disposable all over the globe.

Prior to giving some information about reforms being carried out in our R&D policy, I would like to present some indicators on the state and investment structure of the German R&D system, and on its performance.

The responsibilities for R&D are divided between the federal government and the 16 German states (Länder) which are also mainly responsible for higher education. Government R&D expenditure is roughly shared by both the federal level and the Länder. Research is located in the universities, in the practice-oriented Fachhochschulen (polytechnics) and it is based in large autonomous nonuniversity research institutes like the Helmholtz Centres, Max-Planck-Institutes and Fraunhofer-Institutes as well as in a broad spectrum of smaller institutes co-financed by the federal and the Länder Governments.

Quantitatively the most important factor in R&D, however, is the industrial sector which finances 62 per cent of R&D in Germany. Even with government support 67 per cent of R&D work is carried out by industries, employing 62 per cent of the personnel engaged in this activity. Within the private industrial sector the share of small and medium size enterprises has risen from 12.4 per cent in 1991 to 14.4 per cent in 1995.

For economic reasons R&D investment by industries was reduced in the early 1990s, but recently a new upward trend has emerged. There was stagnation during the last few years in public expenditure due to general budget constraints, however fresh efforts are being made to increase the efficiency of R&D promotion and practice based on the broad political consensus that investment in research and education needs to be intensified both in the medium and long term.

R&D investment by German firms abroad has tremendously risen to 10 billion DM in 1995, i.e. 17 per cent of total R&D investment of industries in Germany. Foreign R&D investment in Germany amounted to 9.5 billion DM in 1995, which means stagnation at a high level.

After a serious debate about Germany's attractiveness as a location for investment in R&D the conclusion drawn was that while the basis was good, and in numerous fields even excellent, in the global competition for technology this knowledge base was not used effectively enough. And the same applies to Europe as a whole.

The comparison of publications in the natural sciences and technology worldwide shows that Western Europe's contribution to primary scientific production is about 35 per cent, while that of the US is about 40 per cent and that of Japan 10 per cent. Europe's share of export of top technological products, however, is only about 27 per cent, while that of the US is about 40 per cent and that of Japan 25 per cent.

So it is not only important to produce knowledge, it must be translated into innovations. In this respect Japan is most successful, while Europe is below average.

At the same time the international flow in the education sector must be kept in mind. The number of students studying in the US, especially students from Asia, has virtually exploded over the last 10–15 years; the number of students from some countries has increased tenfold. By contrast the number of foreign students in Europe has only slightly increased, which is a deficiency. The US attracts the largest number of students and the best minds; over 50 per cent of their PhD students in natural sciences are foreigners. Most of them return to their home countries sooner or later. Those who wish to understand the magnitude of future technology competition must keep their eye on this enormous transfer of knowledge, inspite of the current setbacks in some Asian countries.

Germany and Europe thus face the central challenge of remaining competitive not only in producing knowledge, but also in translating it into practical competence, innovation and thus products and services; and it needs to be transferred to young people from other parts of the world who come to Europe for higher studies and will become its future partners.

I would like to stress that on the other hand partnership and cooperation are equally important worldwide. After all, no one country can now produce all the necessary knowledge alone. The world of global competition, from Germany's perspective requires European and global cooperation.

Initiatives for reform

In order to strengthen the German position in the face of the challenges I have just outlined, German research and technology policy is currently concentrating on three aims:

1. Strengthening Germany's position in knowledge production, specially in key areas of technological development.
2. Translating knowledge into practical competence and thereby into economic innovation.
3. Keeping its R&D system internationally competitive while opening it more than before to other countries and world regions.

With this background in mind, the government has taken a number of initiatives:

STRENGTHENING BASIC RESEARCH

In the 1990s, the federal government and the Länder increased funding inspite of tight budgets for the Max-Planck-Gesellschaft and the German Research Council (the latter is primarily responsible for supporting excellent research in the university sector) by nearly 5 per cent annually. To maintain and sustain this growth in basic research Germany must keep a strong and broad basis.

PROMOTION OF KEY AREAS OF R&D

In recent years, the Federal Ministry of Education and Research has significantly increased support for research in the area of the life sciences, spending nearly one billion DM annually. Germany is moving back to a leading position in this area. The 50 most cited research centres for molecular biology now include 8 German institutes. In 1997, we introduced a major project for human genome research, and funded it with 200 million DM. We launched a competition among institutions so as to foster innovating regions. Industry, research institutes and communities jointly presented their integrated concepts for developing biotechnology in their region. The three regions out of a total of seventeen applicants, who submitted the best ideas received 50 million DM in support for the implementation of their concepts. With this measure alone additional 300 million DM in investments by companies has been elicited, expecting attractive centres of competence to grow as a result there of. In other areas similar initiatives are being undertaken. One such example is a competition for new ideas among young entrepreneurs in the field of multimedia.

COMPETITION IN RESEARCH AND SYSTEMATIC EVALUATION

In future Germany shall introduce more competition via the mechanisms of financing the public sector research centres. For this reason, institutions will have to earn part of their budgets through presenting projects to peer reviews, organized not by the state, but by the German Research Council or the senates of national research centres. Ultimately, the quality of an institute's project proposals will in future determine whether it grows or loses in importance. This effort is expected to enhance motivation and foster greater topicality of research programmes.

To ensure that these research institutions aim at the highest international standards, an evaluation system of comparison with the outside world is being introduced. Thus three of the large research organizations, the German Research Council, the Max-Planck-Gesellschaft and the Fraunhofer Gesellschaft will be subjected to a kind of system evaluation undertaken by independent and international expert groups, something unheard of in traditional German science. The 16 big national centres, now joined in the Helmholtz Society, will be evaluated in a five year cycle. The research institutes cofinanced by the federal government and the Länder are currently being reviewed by the Science Council. At the same time the government is giving more responsibility and flexibility to the institutions within the budget allocation granted to them.

PROMOTING STRATEGIC PROJECTS (LEAD PROJECTS)

The government maintains that science and industry must increase their cooperation in strategic joint projects and that they should jointly define innovative tasks and develop key technologies, novel products and services, to open new markets and employment opportunities or to help in solving social problems. Therefore, it launched in early 1997, after consultations with scientists and industry, a competition for ideas in four scientific and technical areas, and invited scientific institutions and companies to submit interesting and far reaching project proposals in response to the following themes:

— Innovative products based on new technologies.
— Applications of molecular medicine.
— Improvement of mobility in densely populated areas.
— Software for training and lifelong training.

In the first three areas alone, an independent jury selected 42 idea outlines from a total of nearly 500 entries. These received limited funds for development of ideas into concrete project proposals. Finally only the best 3 to 5 proposals in each field will receive matching funds as the lead projects.

In the meantime, three more themes have been announced, again after consultation with high level scientists and industrialists:

— Energy production and storage for decentralized and mobile use.

— Interaction between man and technology in the information society.
— Nutrition—modern methods for food production.

It is expected that with this new method of R&D support futuristic and marketable ideas will be transformed more rapidly into new products and services. Whereas the selection of research fields of strategic importance is done top to bottom, albeit, in consultation with business, industry and the science community, the project partners define their concrete research tasks in a bottom to top way.

IMPROVING TECHNOLOGY TRANSFER AND ENCOURAGING NEW COMPANIES

Germany has a large number of small and medium sized companies and they make a significant contribution to the competitiveness of German industry. These firms play an important role in translating knowledge into practical competence and innovation. For this reason the government wants to increase the transfer of knowledge to this sector and intensify cooperation between them and the world of science for the joint development of innovative technologies and products. That is why it is supporting cooperation among small companies and the science sector. In the last three years, it has approved over 3,000 applications in this special programme, and an increasing number of cooperation projects involve small and medium sized companies and research establishments abroad.

Meanwhile, it is also determined on making venture and risk capital for the establishment of new technology oriented enterprises more easily available, having learnt that in the US over 1,000 companies have been founded in the area of biotechnology, currently employing over 1,00,000 people.

In 1996, within the framework of a special programme, Germany mobilized over 300 million DM in venture capital for new companies in the area of new technologies, this support will continue to grow rapidly in the coming years. Last year, as part of a special loan programme, the government provided low cost loans worth over 1 billion DM aimed at small innovative companies. Within another programme, it supported the founding of 350 companies in eastern Germany, and will continue this support till 1999.

BUILDING CENTRES OF COMPETENCE AND NETWORKS OF COMPETENCE

Along with traditional industrial locations, centres of competence, i.e. clusters of firms and institutions forming a strong potential of complementary knowledge production and innovation are playing a greater role in regional economic development. The example of Silicon valley shows how strongly such a centre can drive development and investments in a comprehensive area of modern technology. There are still other attractive centres of competence in Germany, for example, in areas of laser technology, robotics, production technology, chemistry and automotive technology. We will continue to strengthen such centres and networks in the coming years and will make a concerted effort to establish these as magnets for development and investments, for example, in the area of biotechnology.

DEREGULATION AND INNOVATION FRIENDLY LEGISLATION

Germany has made a major effort to reduce the maze of regulations and bureaucratic procedures in order to improve the framework conditions for R&D and economic innovation. Two examples are:

— Through the amendment of the Genetic Engineering Act (Gentechnikgesetz) of 16 December 1993 and the resulting amendment of relevant ordinances, Germany's legal framework in the area of genetic engineering and its location has been significantly improved.
— The coming into force of the Information and Communi-

cations Service Act in 1997 has given a decisive impetus to the development of new information contents, of innovative applications and the growth of electronic commerce. This 'multimedia act' contains no catalogues of detailed regulations; instead, it gives basic definitions that provide legal clarity and certainty in central questions. This law is the prerequisite, in particular, for the economic use of information networks, within a reliable, secure framework.

PATENT INITIATIVE

The promotion of inventions and of the patent system is an essential part of the innovation process. For this reason the federal government in 1996 started a new patent initiative in order to ensure that research results (including those publicly financed) with a commercial value could be protected and thus made worthwhile. This policy not only created a favourable climate for innovation and patents in research institutions and universities but also a comprehensive and effective patent infrastructure. This approach sensitized scientists to identify marketable research and utilization. In addition, there is a special patent cell for small and medium enterprises to support them in application and utilization of the patent system. Parallel to this the government aims at improving the information framework for SME's access to the European patent system quickly and at low cost.

REFORM OF HIGHER EDUCATION AND MODERNIZATION OF VOCATIONAL TRAINING

A major reform effort targets the modernization of tertiary education. The aims of this reform are to promote competition and differentiation, and to enhance the international competitiveness of German institutions of higher education, through deregulation, performance incentives and orientation towards international cooperation and exchanges.

While in 1997, the federal government and the governments

of the Länder agreed in principle on the contents of a new Higher Education Framework Act, the details of the reform are still under negotiation.[1]

Objectives of this reform are:

- The introduction of performance orientation in the financing of institutions of higher education and at the same time giving them greater freedom in the management of funds.
- Strengthening leadership and management capacities of universities and Fachhochschulen and increasing their accountability.
- Evaluation of research and instruction; involvement of students in the evaluation of teaching.
- Re-definition of periods of study and shortening the duration of courses.
- Introduction of a 'credit point system' for accumulation and transfer of completed studies and tests passed.
- The possibility to award internationally recognized 'bachelor's' and 'master's' degrees.

To enhance the international attractiveness of German universities and Fachhochschulen a pilot programme 'internationally oriented courses of study' was launched in 1996. The funding was meant for developing innovative programmes, especially in the areas of economics, natural sciences and engineering—excellent instruction in the relevant discipline, intensive student counselling, use of a foreign language as a medium of instruction and research with intensive international cooperation and development of internationally comparable qualifications.

Another area requiring modernization is the upgrading of the application oriented institutions, the so-called 'Fachhochschulen', so far catering to about 25 per cent of students and offering shorter courses of study combined with

[1] In Germany education is a Länder subject, the federal government and parliament therefore can only legislate on certain principles, i.e. the framework (editor's note).

internships in industries. These very successful institutions provide young people with well marketable skills and it is the intention of the federal government and the Länder to make the 'Fachhochschulen' even stronger in the differentiated higher education system, and open these to 40 per cent of first year students on the basis of a broader range of subjects which so far concentrated on engineering, business management and social work.

I would also like to mention the efforts being made to further extend and modernize the system of vocational training which to a large extent is organized in a dual form of combining practical training as apprentices in artisans' workshops or in industries with learning in part-time schools. In this way skilled workers are trained who later become masters or technicians and take over responsible positions in firms. The German Government and the social partners, are determined to stick to this practice-oriented training but to adapt it to new qualification requirements. They have undertaken new initiatives to guarantee an apprenticeship place to all school leavers, to modernize training directives according to new qualification requirements and to strengthen life long learning of the workforce.

MONITORING, FORECASTING AND POLICY ADVICE

Some basic instruments have been developed as a basis for an informed R&D policy. In addition to the reports on R&D and R&D policy which the federal government presents regularly, an annual analysis of Germany's technological performance is to be submitted to Cabinet and Parliament. With the help of a group of research institutes an assessment of the weaknesses and strengths of the German system forms a basis for decisions by all interested sectors. The assessment done is internationally comparative, using indicators developed by the OECD and other sources that tell us where we stand and where we are falling behind.

With regard to policy advice and development a new

council for technology was installed directly under the Chancellor, the head of government, in 1995. It convenes leading scientists, entrepreneurs and other outstanding experts to advise on the status of research and technology and new challenges to respond to. The council's first report dealt with the information society, its opportunities and challenges, the second on the state of biotechnology in Germany. The third theme currently under review tackles the issue of 'competence in the global competitive system', with particular regard to requirements for change in the education and training sector, in the structure of industries as well as in the organization of science and research.

Looking farther into the future is the second Delphi-study 1998 on the global development of science and technology. Through a two stage enquiry among more than 1,800 experts from industry, universities, research institutions, associations, and the public service Delphi is intended to give projections of possible future developments with an assessment on questions like: Where does Germany stand on the path to such future visions and how are other partners faring? Reasonably used, this survey also serves as a basis for future oriented decisions in the political field as well as in the science and industrial sector. Three points of general advice by these experts are: To build on international cooperation, to improve our research infrastructure, and to broaden the basis of science and technology through support from sources outside the government.

Internationalization of research and education: Openness for competition and cooperation

Internationalization is an important guideline for German research and education policies. The federal government and the Länder, German companies and research institutions are open to international cooperation. It is well known that competition and cooperation complement rather than exclude each other.

In the process of globalization Germany accepts the challenge as a location for production and services in competition with other countries and regions on the move and offers advantageous conditions to investors, particularly with regard to wage levels and to non-wage labour cost. How can Germany attract foreign enterprises to invest in the country, also in R&D, when more and more German firms go abroad to become active players in the newly emerging markets or are attracted by international centres of competence to which one has to be linked? And finally, how can Germany make itself more attractive to foreign students and young researchers, so that the best, including young people from India, will go for studies not only to the United States or to other countries with highly esteemed universities.

We however need close cooperation, too, for many reasons:

— Science has always been international—excellent research is no longer possible without trans-border cooperation.

— Internationalization of education and training, the openness of its economies and cooperation in the field of R&D are high priority responses to the emerging global knowledge-society.

— A growing agenda of research tasks require international initiatives either because of their worldwide character (environmental protection, climate development, nutrition) or because of their complexity and high costs (genome research, high energy physics or space research).

— Key innovations need for their introduction international standards, they emerge increasingly through cooperation among global industries and in feedback with nationally as well as internationally developed basic knowledge. German industrial managements are aware that they need international partners for reaching breakthroughs and for successfully entering foreign markets. And last, but not least:

— The economic and technological development of the countries at the threshold of industrialization and their integration into world trade has to be promoted by technology transfer and collaboration.

In short, when distances for transportation and communication are shrinking rapidly, a peaceful development of the globe cannot be ensured by competition alone. Cooperation is a necessary component and regulating means.

The government's strategy to strengthen the openness and the cooperation potential of the German R&D and education systems comprises a range of approaches, in addition to classical cooperation agreements:

— Encouraging German research institutions to engage themselves internationally, also by founding branches or even institutes in other countries on reciprocal basis.
— Encouraging German firms to look for partners abroad. Through a special programme the government supports small and medium sized enterprises to build up partnerships not only with research institutes at home but also firms and research institutes in other countries. In this context it started the patent initiative to help SMEs to protect their acquired know-how. This, we think, favours international cooperation because it eases an economic balance of interest.
— Re-thinking the rules of German R&D support in order to facilitate research cooperation that includes foreign enterprises. Firms that have production sites and research labs in Germany and are willing to invest in this country can benefit from its research programmes. And the same applies to science institutions from abroad which participate in German research partnerships. It is understandable that we expect such liberal procedures to be reciprocated by the partner governments.
— We will continue in international research programmes like the human frontier science programme, the international geosphere-biosphere programme, the international human genome project, the world climate research programme

or take part in international space cooperation as well as support to big installations in high energy physics.

— The mobility of students and young researchers to and from Germany is one of the top priorities. Therefore the government is working to make Germany again a place for study and research, as attractive as it had been in the past, i.e. at the time of S.N. Bose. It would appreciate if university partnerships aim at developing integrated study courses that would tremendously increase mobility; mobility, of course, in both directions. It wants to also stimulate more German students and young researchers to go abroad, also to the Asian region, including India.

— Finally, it is also interested in closer international coopera-tion and exchanges in the field of vocational and technical education and training.

In short, the government is convinced, that in times of globalization it would be a wrong strategy for Germany to cut itself off from the outside world in order to protect it from foreign influences. Germany with her tradition of science and industry is dependant on knowledge transfer from and to other parts of the world.

R&D cooperation on the European level and the international cooperation of the European Union

In a political region like Europe, a 'sub-continent' divided into numerous nation states, which has for a long time suffered severe conflicts and wars, international cooperation must, of course, begin at home. Within the European Union, the priorities and structures of a new, the fifth Framework Programme is currently under development. Its guideline is to concentrate the community's support on such joint research efforts that need Europewide cooperation of industries, universities, and other research institutions and on innovative areas which are of decisive importance for economic or societal reasons:

— Reduction of the number of thematic programmes from so far 13 to 4: life sciences, information society, production and transport, and finally, energy and environment.
— Concentration in each programme on a limited number of focal issues or lead themes that are problem or application-oriented.
— And thirdly, promotion of a limited number of key technologies (generic technologies) and the cooperation in the field of infrastructure, thus complementing the thematic priorities. Special promotion of small and medium sized industries, the mobility of young researchers and cooperation with Third World countries will remain prominent topics.

In implementing the programme the relevance of European research promotion shall be assured through advisory groups in which experts and representatives from the world of research, from business and industry and particularly from the consumer's side can participate.

With regard to cooperation with other regions of the world, I think it is very important for European cooperation to become even more open to partnerships with research institutions and enterprises from outside Europe.

Such partnership will be possible in two ways:

Firstly, as hitherto, within the INCO activity (scientific cooperation with Third World countries) out of which to a limited extent European-Indian collaborative projects are or will be sponsored via the ongoing framework programme. To a growing extent the regular thematic programmes will also be open to participants from other world regions which are ready to bear the costs of their project contribution.

The European programme's openness, in principle, to partners worldwide is very much in line with the consensus formulated within the Asia-Europe summit on the importance of intensified science and technology crossflows between Asia and Europe for the strengthening of economic links between the two regions.

Discussions between India and the European Union have just started to identify areas for mutual cooperation under a possible future agreement.

Bilateral cooperation with India and the Asian region in general

In order to intensify cooperation with the Asia-Pacific region Germany has evolved a concept aiming at closer cooperation and more exchanges, including, of course, with India. This is very much in line with the intentions of both governments since the late 1980s, and it was confirmed at the meeting between Chancellor Kohl and the then Prime Minister P.V. Narsimha Rao in 1994 to give bilateral cooperation in research and technology another push and to place it at the centre of relations between the two countries.

Cooperation in this field can build on excellent scientific resources on both sides and on a dense network of bilateral agreements. For this reason Germany is very much in favour of making cooperation a central pillar of the 'Agenda 2000', a political project that defines Indo-German relations for the year 2000 and beyond in a comprehensiveway. There are already clear areas of priority in this cooperation: biotechnology and genetic engineering, medical and health research, new materials, information technology and software development. A particular focus is on space technologies and their joint applications for civil purposes through close cooperation between the German Centre for Aeronautical and Space Research and the Indian Space Research Centre.

Exchanges of students and young researchers are essential for underpinning closer cooperation in the future. Institutional partnership should be actively used to stimulate mobility. They are particularly appropriate to further the knowledge for the partner's strength in teaching and research and to create a framework and practical routes for regular exchanges. Such training and research partnerships will lead to more exchanges in both directions, also from German universities and

Fachhochschulen to India. An outstanding new programme for promoting mobility is the 'sandwich' exchange programme in engineering sciences allowing Indian graduate students from the five IITs to do research for their master theses at five German technical universities and to spend training periods in German firms.

Finally, both sides agree that enterprises, including SMEs with research potential or interest should be involved more actively in joint research projects and that such projects should be oriented more towards industrial needs and economic innovation. In making step by step progress both countries shall certainly help each other on the way to successful economic development and a better employment situation, which is a common concern.

Plate 9: Jawaharlal Nehru and Albert Einstein shared a concern for non-violence and world peace. The Indian Prime Minister met Einstein at Princeton in 1949.
Courtesy: Nehru Memorial Museum and Library, New Delhi

Plate 10: In the 1920s Berlin was a vibrant centre of discoveries in physics. Five Nobel laureates met here in 1931 (*from left*) Walther Nernst, Albert Einstein, Max Planck, Robert A. Millikan, Max von Laue. Courtesy: Stiftung Preussischer Kulturbesitz, Berlin

Plate 11: Ongoing friendship: S.N. Bose lectured at the Fritz-Haber-Institut in Berlin in 1953 at the invitation of Prof. Max von Laue.
Courtesy: Archiv zur Geschichte der Max-Planck-Gesellschaft, Berlin-Dahlem

Plate 12: Prof. Reimar Lüst, the President of the Alexander-von-Humboldt-Stiftung, fascinated the audience with his 9th Bose-Einstein Lecture on Comets. From left: Ambassador H.-D. Dieckmann, Prof. A.M. Khusro, Prof. R. Lüst, Stephan F. von Welck, the editor.
Photo: S.K.Sharma

From Bose-Einstein to 2+2 Projects: 75 Years of Indo-German Cooperation in Science and Technology

PETER ENGELMANN

Introduction

Science has always been universal. Attempts to limit science to individual nations have failed and have led to a setback of scientific progress in that very nation trying to restrict or control science. Scientists want to discuss their ideas, their problems and results with colleagues working in the same field, irrespective of their nationality. They need the response, be it confirmation or critique, and new ideas for their future work. New ideas come more often from colleagues working in other laboratories than from their co-workers at home. Without this response or new ideas scientists will lose their creativity. It is for this reason that scientists often identify more strongly with members of the international scientific community than with their own nationals—at least as far as their own research work is concerned.

Cooperation, as also bilateral cooperation in science, in principle is part of the normal life of the international community of scholars. It can be stimulated by providing funds for the exchange of researchers for bilateral cooperation, guided in certain directions by the flow of money. Pure scientific cooperation works most efficiently where bureaucracy is kept to a minimum.

Indo-German scientific cooperation responded to these attitudes of the science community. The government and some

*Fifth Bose-Einstein Lecture, 5 November 1997, India International Centre, New Delhi.

foundations provided money to set up an appropriate organizational structure consisting of very few people who could act freely within certain limits. The result was a lively cooperation which started without great delay and which produced good scientific results as well as many personal friendships among the scientists of both countries. Coordinators saw to it that the joint projects were oriented towards fields of common interest in tune with the intentions of the governments and the overall goals of their research organizations.

Technology is based on science. It is the link between scientific results and their application for the benefit of the people. Technological development is not quite the same as scientific research. It is application oriented and has the market in mind. Technological leadership means leadership in certain market segments, economic strength and also power. For these reasons governments and commercial companies take greater interest in technological development. Scientists and engineers working on the development of new or improved technologies have never been as free in the exchange and publication of their ideas and results as their colleagues in pure science. Technological know-how means money and for this reason has to be protected to some degree. The borders between science and technology are fluid, restrictive handling of technological results might be counter-productive, because it may prevent the necessary feed-back and improvements, which could lead to a wider market.

Cooperation in technological development requires certain rules to be followed. In bilateral cooperation the government and/or appropriate organizations of both countries have to conclude agreements defining the rules, how technological developments are to be handled and eventually used commercially. Without clearly defined rules, which meet the interests of both the parties involved, there will be no cooperation. The closer technological development comes to new products for the market, the more questions of intellectual property rights and economic structures of the two countries come into play. For several decades, the diverging economic systems of Germany and India have prevented a close co-operation in product oriented technological development. This is one of the reasons why in the past most of this bilateral cooperation was in basic and applied science rather than in technology. Since several years, however, India is on the move towards a new economic policy, close to the German economic order. Accordingly, cooperation enters into a new phase and is changing its character.

Both governments wish to widen cooperation, to improve industry and to increase benefits for both economies and people. Therefore, they have recently set up a 'Committee on Science and Technology' to speed up the reorientation of bilateral cooperation and to think about ways to stimulate the interest of industry in joint technological projects. In the first meeting of this committee held in 1996 ways and means have been discussed, how the existing obstacles could be over come and whether the organizational structures of the actual bilateral cooperation have to be changed. This process is still under way, but the character of cooperation is already changing and in a growing number of joint projects either German or Indian industry or both could be included.

The early scientific contacts, Bose and Einstein

Let me go back to the early scientific contacts between the two countries, to the cooperation between S.N. Bose and Albert Einstein. To understand this phase, one has to look at the early decades of this century. At that time Europe dominated the world also in science, and India was still a British colony. There already existed a transnational scientific community which interacted closely. The communication between scientists was not quite as fast as today in the age of e-mail, but it worked and the distances within Europe were relatively small. Physics and chemistry saw a revolution: Age old principles of understanding the world and its phenomena like Newtonian physics and Euclidic geometry could no longer describe the behaviour of the macro- and microcosmos. New methods and laws had to be found.

In 1900 Max Planck discovered his famous Radiation Law and laid the foundation of the Quantum Theory. Soon after that Albert Einstein produced papers on the light-quanta or photon and on the photo-electric effect and developed the 'Special Theory of Relativity' linking matter and energy. In 1916 followed his paper on the 'General Theory of Relativity' and in 1919 Einstein's hypothesis on the bending of light by strong masses could be proven during a total solar eclipse in South America. Space, time, and matter or energy from now on had to be considered as one physical reality. The whole picture of the world had changed.

Berlin, where Einstein lived, was then a central place in scientific development. The German language was a must for physicists at that time, as is English today. The important research papers were published in the *Zeitschrift für Physik* or in the *Annalen der Physik*, both of which accepted only contributions in German.

This was the science scene when Satyendra Nath Bose grew up and studied mixed mathematics at Calcutta University. Born in 1894, he graduated in 1915 and became a lecturer in physics in 1917. In his early twenties he published papers on topics of theoretical physics, some of them together with Meghnad Saha. Both of them studied German and in 1919 translated Einstein's original papers on relativity into English. In 1921 Bose joined the new university at Dhaka as a reader in the Department of Physics. At the age of 30 he produced his paper 'Planck's Law and the Hypothesis of Light Quanta' which was to become famous and which started his cooperation with Einstein. On 4 June 1924 he sent this paper to Einstein, the 15 years older Nobel laureate, asking him for his opinion and his help for its publication in the *Zeitschrift für Physik*. On 2 July, only four weeks later, Einstein sent a post-card (see p. 27) to his young colleague telling him that he considered his paper a beautiful step forward and that he had translated it for publication. The *Zeitschrift für Physik* printed it in its September issue.

Many things worked rather fast at that time! Einstein did not hesitate to read a paper of a still unknown young Indian and immediately understood the deeper significance of Bose's derivation of Planck's law, the new concept of Bose Statistics. Einstein applied it to the case of ideal gases, predicting the existence of 'Bose-Einstein condensation', which could only recently be experimentally verified. This long delay is due to the enormous experimental difficulties to cool down an atomic gas to a temperature of only 1 to 2 Mikrokelvin without solidification. Wolfgang Ketterle and his group at the Massachusetts Institute of Technology have now successfully demonstrated coherent behaviour in Bose-Einstein condensates made from sodium atoms. The condensates are millimetre size clouds of very dilute sodium gas with a density of only 10^{14} atoms per cm^3. Bose condensates form a source of coherent atoms for experiments in atom optics and for the development of so-called atomic lasers. The MIT group is now developing an output coupler on the basis of magnetic mirrors to extract the coherent matter waves. Such beams would have a lot of applications, as the matter waves would have a lower wavelength than optical lasers and could also easily be focused to a point. This opens up new possibilities for the fabrication of integrated circuits with even smaller dimensions. This so-called nano-technology could revolutionize the computer technology of today. It is amazing to see that 75 years after the prediction of the phenomenon by Einstein, the practical applications of Bose's work are not yet exhausted.

In 1924 Bose was granted two years' leave to travel to Europe at the terms of continued pay by Dhaka University. He received the salary as an advance payment of Rs. 13,800. Bose went first to Paris and in October 1925 to Einstein in Berlin, where he met many eminent scientists of the time. Max Planck, Peter Debye, Fritz Haber, Max Born, Otto Hahn, Lise Meitner, Walter Bothe, Max von Laue, Wolfgang Pauli,

Werner Heisenberg and many others who were working and teaching there. In addition to his scientific research, Bose used his stay to attend seminars and colloquia, and cultivated personal contacts with many scientists discussing problems and new ideas with them. So he exercised, what guest scientists still do today, that is, scientific exchange and co-operation.

It was not a cooperation covered by any governmental agreement or by agreements between research organizations of two countries. Nevertheless this scientific exchange on a person to person basis achieved the desired results, which is still the aim today but as a more formally organized co-operation: Bose's long term stay at German laboratories and his close contacts with German scientists shaped his way of thinking that determined his further scientific career. After returning to Dhaka he began to reorganize the Physics Department, making use of his experience. He launched, among other things, activities in experimental physics, an area which was not yet well known in India.

The contact between Bose and Einstein was followed by many others in the late 1920s and early 1930s: Sommerfeld visited India in 1928, he met Bose and stayed with Raman in his institute in Calcutta, where von Karmann and Heisenberg also spent some time as visitors. Raman and Sommerfeld became good friends and Raman went to see Sommerfeld in 1930 in Munich. In 1935 Max Born came to the Indian Institute of Science, where Raman was then working and stayed for six months, being a valuable teacher for the young Indian students.

The German Academic Exchange Service (DAAD), existed already in the 1930s and supported the international exchange of scientists and provided scholarships for foreign scientists coming to Germany. It seems, however, that an exchange between Germany and colonial India had not been part of the DAAD activities at that time.

The golden period of science in Germany came to an abrupt end after Adolf Hitler seized power. Many reputed scientists left the country or had to flee in order to escape the persecution of the Jews. Einstein was one of the first to leave Germany. Many Jewish scientists went to the USA and contributed to the growing importance of that country in science. The first era of scientific cooperation between India and Germany ended in the 1930s. The contacts that still existed were definitely interrupted with the beginning of World War II.

Early post-War cooperation: exchange of scientists and projects for the improvement of the scientific infrastructure in India

After the War the world underwent dramatic changes: India became independent and divided in 1947. Germany was occupied. In 1949 two separate German states were established, the Federal Republic of Germany (FRG) and the German Democratic Republic (GDR). The world split into East and West, somewhat later the Third World entered world politics.

All these changes had an effect on science and technology, especially on scientific cooperation.

In India Jawaharlal Nehru, the first Prime Minister, strongly supported science, which he considered an important basis for the progress of the country. Therefore existing research organizations like the Indian Council of Scientific and Industrial Research (ICSIR) were enlarged, universities got expanded, new research organizations and a new type of high ranking technical universities, the IITs, were established. By these measures the basis for scientific and technological cooperation of India with industrialized countries also became stronger.

In Germany most of the universities and research laboratories had been destroyed and many young scientists had been killed in the War. It took a number of years until scientific work could be fully resumed. In the FRG the older scientists tried to link up with the organizational structures of the pre-Nazi period and to reinstall autonomous scientific organizations. Having experienced the absolute control and nationalistic orientation of science by the Nazi rulers, the scientific community was eager to reduce the influence of the government to a minimum. The re-establishment of traditional organizations like the Max-Planck-Gesellschaft (the former Kaiser-Wilhelm-Gesellschaft) and the principle of scientific self-government helped regain part of the lost reputation of German science. In order to speed up the reintegration of Germans into the international scientific community, contacts and exchanges of scientists were strongly supported. In 1950 the DAAD, in 1951 the German Research Council (DFG), and in 1953 the Alexander-von-Humboldt Foundation (AvH), were re-established; three organizations which very actively supported international scientific co-operation.

In the 1950s and 1960s a new phase of cooperation between India and the Federal Republic of Germany started. It was mainly based on (a) AvH scholarships for young Indian postgraduate scientists, (b) on the exchange programme of the DAAD for German and Indian students, and (c) on the participation of Indian scientists in projects sponsored by the DFG. To organize the cooperation, agreements were concluded between the University Grants Commission (UGC) and AvH and DAAD respectively, between the ICSIR and the DAAD, and between the Indian National Science Academy (INSA) and the DFG. Since the end of World War II, the AvH has awarded 1,110 scholarships to Indian scientists, while the DAAD has awarded 1,610 long-term and 1,655 short-term scholarships for Indian students going to Germany and an equal number of stipends for German students coming to India.

The Indo-German co-operation reached a new dimension when the German Government decided to provide support for projects strengthening the scientific infrastructure of India. The Indian Institute of Technology (IIT) at Madras was the first and outstanding example of such a project, but many other followed like the Research Vessel Sagar Kanya and special laboratories at ICSIR institutes and at universities like the Anna University at Madras.

The German support for these projects through the Ministry of Economic Cooperation (BMZ) was not limited to funding. In the case of the IIT Madras, many qualified German professors spent years at the institute teaching, building up laboratories and workshops, and helping in developing new curricula. Professors and lecturers of the IIT, on the other hand, came to German universities for training and scientific exchange for extended periods of time. A great number of joint research projects of the IIT Madras could subsequently be carried out with German partners funded by the BMZ.

Thanks to this major cooperation here and in numerous other research projects funded by the BMZ many personal friendships and also close ties between the various departments of the IIT Madras and German universities like Karlsruhe and Aachen, developed which even today form the basis of many bilateral scientific contacts and projects sponsored on the German side by the Ministry of Education and Research, (BMBF). In a similar way, the R.V. Sagar Kanya played and still plays a major role in the project cooperation in marine science.

Agreements and cooperation in scientific and technological projects

A major step forward in scientific cooperation was made, when in the early seventies the Governments of India and of the Federal Republic of Germany concluded two agreements: In 1971 the 'Agreement on Cooperation Regarding the Peaceful Uses of Atomic Energy and Space Research' and in 1974 the 'Agreement on Cooperation in the Field of Scientific Research and Technological Development'.

Under the umbrella of these two agreements a series of 'Special Arrangements' were subsequently concluded between research organizations or institutions acting as nodal agencies for the cooperation in the two countries. Through these special arrangements efficient procedures were defined for the implementation of joint research projects and the exchange of scientists. On the German side the Federal Ministry of Education and Research (BMBF), provided means for funding travel, assignments and special equipment. Coordinators and international officers were appointed for the definition and execution of the bilateral programmes.

In support of this cooperation both countries established the position of counsellors in science and technology in their embassies in Bonn and New Delhi respectively in the mid-1980s. These counsellors have since proved very valuable in creating new contacts and sorting out many problems on spot.

The number of bilateral projects grew very fast, from about 20 in 1974 to about 130 in the early 1990s. The exchange of scientists within these projects soon reached a level of 180 to 200 per year. A joint project is a well defined scientific or technological task to be solved in a balanced manner by groups of scientists in India and Germany over a period of about three years. The exchange of scientists allows their integration into a team of the partner country, the exchange of ideas, the transfer of methods and know-how, and the joint use of major experimental facilities. In some cases special equipment was and still is transferred to the partner laboratory to enable it to contribute better to the common goal.

Many of the projects were proposed by scientists of both countries, who had already been in contact. Expert missions, joint seminars and workshops were organized to open up new fields of cooperation or, later on, to review the results of joint projects in a certain field and to decide on the directions of future work. By now about 75 Indo-German workshops and seminars have been held.

The cooperation covers almost all fields of science. The focal areas have been adjusted with time according to the actual interests of the two countries. In the beginning the cooperation between the Department of Atomic Energy and the Forschungszentrum Jülich (FZJ), on topics of nuclear reactor technology and safety formed a major part of the bilateral activities. However, different reactor lines developed in the two countries with diverging policies of Germany and India towards the Non-Proliferation Treaty, due to which the cooperation had to be restricted to generic and non-sensitive areas. Several years ago, projects on reactor technology and its safety were phased out completely. In the cooperation of the Deutsches Zentrum für Luft- und Raumfahrt (DLR), (German Aeronautics and Space Centre) and the Indian Space Research Organization (ISRO), on space research, the second topic of the first governmental agreement, technologies with potential military applications, had to be excluded for similar reasons.

In general, a bottom to top approach has been chosen in defining joint projects in order to guarantee a high degree of motivation of the scientists. Overall guidance could nevertheless be achieved by a proper selection of project proposals and by the organization of workshops in desired areas. Some of the major areas of project cooperation were and are:

- Aeronautics and space research
- Medical research
- Life sciences and biotechnology
- Energy research
- Marine and geosciences
- Materials research, metallurgy and surface engineering
- Microelectronics and information technologies
- Environmental research, and
- Fundamental research in physics.

Project partners come from all kinds of research organizations and from many universities of the two countries. On the Indian side, the ICSIR institutes, the research centres of the Department of Atomic Energy, of ISRO and the Indian Council for Medical Research (ICMR), and the IITs deserve special mention. On the German side many of the Helmholtz centres like FZJ Jülich, Forschungszentrum für Gesundheit und Umwelt (GSF), München, Deutsches Krebsforschungszentrum (DKFZ), Heidelberg and DLR Köln, but also some institutes of the Max-Planck-Gesellschaft and of the Fraunhofer-Gesellschaft, many universities, hospitals and some private firms were and are partners of joint projects.

About half of the projects pertained to basic science, while the other half were application oriented. The majority of them were rather small in nature and needed only additional funding for the exchange of scientists. Some, especially in areas of energy research, biotechnology, marine science, geology, surface engineering, and space research received special funding for expensive equipment, collection of data, construction and testing of prototypes, the use of research vessels and space vehicles, etc.

Measurable results of the project cooperation are more than 1,000 joint scientific publications and about 10 patents or patent applications. To mention only a few:

The long-term cooperation of the National Institute of Oceanography (NIO) in Goa and Hamburg University on the interdependence of atmospheric and oceanic events has led to important contributions to climate research and to the International Geosphere-Biosphere Programme.

The cooperation between ISRO and DLR in space research included the development of a special camera, the Multispectral Optoelectronic Scanner (MOS), which was launched in 1996 into a polar orbit by the Indian satellite IRS-P3 for remote sensing of properties of the atmosphere and the ocean. The data already obtained are of such international interest, that agreements could be signed with the European Space Agency (ESA), and NASA on their common use. A combination of marine research and remote sensing from space has recently been started, promising a wealth of information.

Joint work on geology is yielding insights into the development of earthquakes on the Indian shelf and on the tectonic plate movements. A major joint project in the Koyna area in west India will improve our knowledge about the physical processes involved and the role of fluids in earthquake mechanics. An array of 21 scientific wells is being installed and equipped with sensors for a five year monitoring programme making use of the seismicity induced by the varying stresses generated by the level changes of the Koyna dam reservoir, which appears to be the highest in the world. These examples show that the projects are basic in character. They lead, however, to results of practical use for society.

Projects have also yielded more direct applicable results of commercial interest, as can be seen from joint patents applications. These are mainly in areas of biotechnology, pharmacology, and medical technology. I hope that the still relatively smaller number of joint patents will increase with the change of the character of bilateral projects.

Not measurable are the results of the exchange of ideas for mutual benefit and personal friendships. A review I made among the German project partners in 1995 showed the following:

- The project partners judge their cooperation as valuable and worthwhile
- Personal knowledge and common scientific interests are the key condition for success
- The cooperation widens the research possibilities and generates new ideas
- Learning from different cultures and ways of approaching problems has a fertilizing effect.

Almost all German project partners were eager to continue cooperation with Indian counterparts.

In addition to the project cooperation sponsored by the governments, there also exists a project cooperation financed by private sources. The Volkswagen Foundation, for example, has supported 57 application oriented Indo-German projects since 1979. Partners come from German universities and from IITs, universities, ICSIR and ICMR laboratories in India. Projects and exchange of scientists also exist in the framework of partnerships between Indian and German universities.

The intensity of cooperation and usefulness of the joint projects for scientific progress, for solving general problems of the people, and for strengthening the relations between India and Germany has always been noted with satisfaction by the two governments. In recent years, however, concern was expressed about the low involvement of industry and the low number of projects leading to industrial innovations. The two governments, therefore, took steps to explore possibilities for a stronger orientation of the project cooperation to industrial needs.

New approaches and perspectives for future cooperation

In 1988 Chancellor Kohl and the late Prime Minister Rajiv Gandhi confirmed their mutual interest in intensifying cooperation in science and technology. An 'evaluation committee' for such Indo-German cooperation was set up that year to review the ongoing activities and to work out recommendations for future cooperation. This group of high ranking and experienced experts explored with great effort the possibilities for increased industrial R&D cooperation and reported their findings towards the end of 1990 and early 1991 respectively to their governments. Most of the recommendations, unfortunately, could not be taken up because they came at the wrong time: In Germany reunification had just taken place and absorbed all energies, while in India, the liberalization process had not yet started. Both factors lowered the interest of industry in cooperation. The manifold contacts between representatives of German and Indian industry established during the evaluation process, however, have been useful for the interindustrial relations which developed after the beginning of liberalization and for the formation of joint venture companies.

A second step to intensify and widen the bilateral cooperation was made by the two governments in the mid-1990s by setting up a 'German-Indian Committee on Science and Technology'. The committee held its first meeting in Bonn in October 1996 under the joint chairmanship of the German Ministry of Education and Research (BMBF) and the Indian Department of Science and Technology (DST). It initiated measures to improve the flow of information about the different bilateral activities and to increase the involvement of industry. It suggested to work together in so-called 2+2 Projects, in which research organizations and industry jointly participate on each side. A working group was established to implement these proposals and to work out recommendations for future actions.

The character of our cooperation is already changing. Industrial partners are being included in a growing number of projects and a stronger orientation of new projects towards marketable products is underway. It will take, however, great efforts on both sides to reorient the bilateral programmes in such a way that the opportunities which have been provided by the scientific and technical relations established in the successful cooperation of the past can be fully exploited.

Let me summarize and give my personal views on the perspectives of future cooperation:

- The cooperation in science between India and Germany has a long tradition. It started in the early decades of this century on a person to person basis.
- Soon after World War II and after India had become independent the cooperation was more formally institutionalized and expanded. It started with an exchange of scientists. German assistance in improving scientific infrastructure in India followed as a second step.
- In the early 1970s, two inter-governmental agreements were concluded. They formed the basis for an intensive and fruitful cooperation in scientific and technological projects.
- These projects, that led to good scientific results and built up a wealth of excellent relations between German and

Indian scientists or scientific and educational institutions, are now being supplemented by projects oriented towards industrial innovations and marketable products.
- Industry is expected to participate in a growing number of joint projects with the aim to establish direct R&D relations between companies of the two countries.

The bilateral cooperation depends very much on the individuals involved. Common interests, mutual trust and understanding are preconditions for success and have been achieved among the scientists. In order to widen the co-operation to industry, new foundations have to be laid. I believe that this can best be achieved by a supplementary programme for the long term exchange of young graduates and students, who not only spend some time studying at the universities of the partner country but also spend extended periods for practical training at industrial firms. Two such programmes were under discussion and at least one of them already started in 1998. Scientists and engineers who have been engaged in such programmes, in their future jobs in the industry will be oriented towards the partner country. It is hoped that they will understand the culture, the ways of thinking and working in the host country and therefore be able to better overcome problems in bilateral cooperation in areas of commercial interest.

Other preconditions for a stronger involvement of industry in R&D cooperation are:

- a similar economic system in the two countries
- the existence of a sufficient number of in-house R&D units of equal standing in industrial firms
- the awareness of the potential benefits of cooperation with the foreign partner
- trust in the legal conditions and industrial property regulations of the partner country, and
- the reduction of bureaucratic hurdles.

As in any area, in R&D cooperation with a foreign partner, industry will and must take its own decisions based on sound commercial considerations. In the long run, industry must take the lead and initiative in this type of cooperation. The governments can help to establish adequate conditions and see to it that the preconditions will be fulfilled. It can also help to bring potential partners together via the 2+2 Projects, in which they can learn to work together without taking a major risk. It is my opinion, that the governments should, however, not try to influence the decisions of industry when granting financial support. Joint projects carried out by an industrial firm only because they get funds from the governments normally do not lead to the kind of long term relations both countries want to build up.

I am sure it will be possible to add a similar successful cooperation in the development of technologies and products based on the past successful scientific cooperation. This will not come automatically or immediately, because new players will be involved and thus both countries can only make limited use of the mutual trust established during the decades of scientific cooperation. If they work in the right direction they will, however, succeed.

first on the 28th last Heisenburg spoken in the Colloquium about his theory, then, in the last Colloquium, there was a long lecture on the recent hypothesis of the spinning election, (perhaps you have heard of it) Everybody is quite bewildered, and then there is going to be very soon a discussion of Schrödinger's papers. Einstein seems quite excited about it; the other day coming from the Colloquium, we suddenly found him, jumping in the same compartment, where we were, and forthwith he began to talk excitedly about the things we have just heard. He has to admit that it seems a tremendous thing, considering the lot of things which these new theories correlate and explain, but he is very much troubled by the unreasonableness, of it all. We were all silent, but he talked almost all the time, unconscious of the interest and wonder that he is exciting in the mind of the other passengers.

3. From Bose's letter to Jaqueline Eisenmann in Paris, 1926

Daimler-Benz Research and Technology: New Strategies for a Global Market

KLAUS DIETER VÖHRINGER

In the town of Ulm in southern Germany, where Albert Einstein was born, a statue stands in front of the railway station. It comes from India. A plaque attached to the statue bears the legend: 'Dedicated to the City of Ulm by the Indian People'. The names of Satyendra Nath Bose and Albert Einstein stand for excellent scientific achievements in this century. But they also stand for the long tradition of collaboration in the field of science and research between India and Germany.

The Daimler-Benz AG intends to continue this tradition. In 1997, it opened its first research centre in India. This too, was born in Ulm, where Daimler-Benz has located its research in the field of information technology. The Daimler-Benz Research Centre India is an important element of the offensive launched by the company at the beginning of this decade to internationalize its research activities. This offensive is based on far reaching strategic plans.

Daimler-Benz: fields of business and research activities

Transport, traffic and services—are the three strategic orientations of the Daimler-Benz group. Today it operates 23 business units which can be grouped under the following broad headings:

*Third Bose-Einstein Lecture, 2 May 1997, India International Centre, New Delhi.

- Cars
- Commercial vehicles
- Aerospace
- Services
- Rail and
- Microelectronics

A sales volume of over 100 billion DM makes Daimler-Benz the biggest industrial company in Germany and one of the largest in the world. A workforce of 2,90,000 people world-wide produced an operating profit of 2.4 billion DM in 1996.

The top performer is the vehicle business which accounts for around three quarters of group turnover. Last year, sales figures in both the commercial vehicle and the car divisions reached new records in the company's history.

Group research comprises the entire research activities in all its business units. The work of this sector can be summed up under three broad headings:

- Firstly, securing the technological future by means of an innovative group research function, oriented to its business strategies and business units.
- Secondly, carrying out group-wide technology management in conjunction with the entrepreneurially independent business units.
- And thirdly, guaranteeing the Board's technological expertise.

The main focuses of their research lie in the following spheres:

- Man and vehicle
- Vehicle systems
- Methods and tools
- Drive systems
- Energy technology
- Traffic and traffic environment
- Microelectronics and microsystems

- Automation and drive systems
- Information technology
- Products research and environment
- Materials and
- Social impacts of technology.

Around 1,500 employees work in these areas at six different locations within Germany.

Two highlights of such research clearly underscore the technological competence and innovative capacity of its enterprise. One such concept study is called 'F200 Imagination' for a coupé for 'the day after tomorrow'.

It combines a great number of innovative systems to enhance safety and comfort. For example, the company implemented a forward-pointing 'drive-by-wire' system in the F200. It allows the driver to control all movements of the car with so called side-sticks integrated in the interior lining of the doors and the centre console. Since it eliminates the conventional operating controls, steering wheel and pedals, it opens up completely new opportunities for interior design.

The most conspicuous exterior features are the hydraulic pivoting doors that make for comfortable entry and exit even in tight parking spaces. The F200 also has special safety features to offer. Should a crash occur, in addition to the familiar side airbags a so-called window bag unfolds out of the roof frame like a curtain to protect the occupants on both the front and back seats. With these and a number of other novelties, the F200 contains a wealth of innovations for the consumers in top Mercedes models of 'the day after tomorrow'.

To lay the technical foundations today for the environmentally compatible mobility of tomorrow—is the maxim of the research work on alternative drives and sources of energy. The group sees particular potential in fuel cell drive. It utilizes the advantages stemming from the principle of electric drive and at the same time avoids the drawbacks of a battery powered electric vehicle.

In 1997, the company presented the first fuel cell powered multi-purpose vehicle which can operate under everyday conditions. After intensive research, it succeeded in increasing the power density of the fuel cells, thus reducing the weight and volume to such an extent that the entire fuel cell system can be accommodated under the rear bench seat.

In its first test vehicle in 1995, the fuel cell system filled almost the entire volume of a van. Today, its size has been reduced to such an extent that six people can travel in the fuel cell powered people carrier. This demonstrates the dramatic progress made in this period.

Challenge of the global market

Following this exemplary look at the research activities, I would now like to come to the main subject—the strategies with which Daimler-Benz research meets the challenges of the global market. The question already contains a very essential premise:

Its market is the world market. This is a trivial statement, nevertheless, it is more relevant today than ever before. Globalization of economic relations takes place on the macroeconomic level as well as on the level of the enterprise. It is based on a large number of factors and is constantly gaining in dynamics.

Changed political parameters such as the establishment of free trade zones or the political opening and economic liberalization of former socialist countries has given this trend additional impetus. But new technologies in the fields of transportation and communication also have the effect that consumers and companies today have access to an international supply of goods and services practically worldwide.

Until a few years ago production facilities, and technical and organizational know-how were the reserve of the classic industrial countries. Today they are available worldwide. This globalization of supply is accompanied by a rapid growth in the worldwide flows of capital. Investments are made where the markets are and where the prospects for returns are the best.

At the same time, dramatic changes in the weightage of various economic regions can be observed. The triangle of the USA, Japan and Western Europe still dominates the development of the world economy, but it will gradually be superseded by a far more complex structure which includes the newly industrializing countries of Asia and South America and the reform countries of Eastern Europe.

The idea of developing and producing a product in one place and delivering it to the entire world is no longer practical today. The needs of customers are becoming more and more differentiated. Consequently it becomes more and more important to offer custom made products, tailored to the individual and regionally differing desires of customers.

For this purpose the entire value-added chain of the company must be globally oriented starting from the area of research and development, continuing through purchasing, production and sales, all the way to financing and human resources policy.

Globalization in the Daimler-Benz group

Globalization has a long tradition in the Daimler-Benz group. Already towards the end of the 1970s, the Mercedes-Benz brand was represented with its vehicles in more than 160 countries of the world. No motor vehicle manufacturer has a broader and more efficient worldwide sales and service network than Mercedes-Benz. The internationalization of production was first promoted in the commercial vehicles division.

The group availed the advantages of an interlinked international production system early on to secure for itself world leadership in the area of trucks over 6 tonnes and buses over 8 tonnes. As early as the beginning of the 1950s it set up its own production plants in Brazil and Argentina. A further milestone on the way to becoming a global commercial vehicle manufacturer was the construction of an assembly plant here in India together with the Tata group. In 1981, it acquired the American truck makers—Freightliner—a very successful bridgehead in the world's largest market for commercial vehicles, the USA. Today, Daimler-Benz already produces every second commercial vehicle abroad.

In the passenger car segment as well, in which the company operates as premium suppliers of high quality cars, a change is in the making, from 'Made in Germany' to 'Made by Daimler-Benz'.

The decisive reasons for this development were not merely the opportunities perceived for its products in the promising markets of Asia and Latin America, but rather the fact that the group decided, within the framework of an extensive product offensive, to expand its traditional range and tap new growth markets and market segments with attractive products.

The new A-class typifies this. Shorter than a compact car, it offers the spaciousness of a medium sized car, the comfort of a saloon and the practical features of a minivan, while for the first time introducing Mercedes-Benz's exacting safety standards into the subcompact car segment. In addition to being built in Germany, the A-class will also be produced in Brazil.

Further examples are the V-class people carrier, in production in Spain since 1996; or the M-class, a new off road vehicle, for which a new factory was built in Tuscaloosa in the USA. And then there is the Smart, specifically developed for operation in congested urban areas, the two seater city runabout which rolled off the assembly line in France early in 1998.

But the increasing internationalization includes all other business units of the group as well. Daimler-Benz Aerospace, which houses activities in the fields of aviation, space, defence and drive systems, makes three quarters of its turnover in international cooperation, for example, in Airbus Industries. Daimler-Benz Interservices, founded in 1990 as a group service-provider already has 166 branches in 28 countries today. The group subsidiary handling microelectronic products and automotive equipment, Temic, has development, production and sales establishments in 37 countries worldwide. And our railway offshoot Adtranz, a 50:50 venture with Asea Brown Boveri, operates all over the world from Australia to Asia, Europe to America and Africa.

Reasons for internationalizing industrial research

Against this backdrop it is clear that the research departments of the company too must critically examine whether their structures adequately take into account the changes in the international markets which have already occurred and are expected in the future. Yet, when the group began in the early 1990s to apply itself intensively to the problem of the internationalization of its research activities, it discovered that this question was often viewed with mixed feelings around the world.

The internationalization of research has increased sharply over the last one and a half decades. In the first half of the 1980s, enterprises began supporting their local branches abroad with capacities particularly in the area of adaptive development. The objective was to adapt technologies developed in the parent country to local production and sales requirements. Since the mid-1980s there has been a clear trend towards strengthening and expanding the capabilities of foreign R&D. Today, high calibre research is increasingly carried out at foreign locations.

However, the process of internationalization in the 1990s goes hand in hand with an increasingly selective choice of R&D locations and a concentration on a few leading centres. Frequently only one competence centre is established worldwide per product group or field of technology.

These measures can be interpreted as a reaction to certain 'uncontrolled' growth in the 1980s. Then not only the number of R&D laboratories distributed around the world grew, the problems grew simultaneously: Duplication of activities, subcritically equipped institutes and demarcation disputes are just a few of them. At some companies this even led to a recentralization of research and development in the home country.

The following quote is taken from a study of the late 1980s. I think it expresses vividly and in a nutshell the mixed feelings in many enterprises: 'Globalization of R&D is accepted more with resignation than with pleasure.... In one of our own cases internationalization was almost described as an unavoidable nightmare, closer to a marketing gimmick than to an effectively contributing R&D outlet.'

Even today, there is still no 'magic recipe' for the 'right' amount of internationalization of industrial research. This is the outcome of a survey of 22 renowned globally operating companies, which was conducted last year in Japan, the USA and Europe. As to the degree of internationalization, two distinct groups of enterprises crystallize:

Nine of the reference companies concentrate their entire central research at a single location, always in the home market.

Twelve enterprises have distributed their research among several locations, with a definite 'home base'. The scale of international research activities varies appreciably, though. The research institutes abroad number between one and eight; the number of scientists employed there between 20 and 1,200. Five of the firms declared their intention to establish one or more additional research facilities abroad.

Apart from the question 'how much?' the survey also asked 'why?'; that is, what are the motives of companies for investing in the build up of research capacities outside their home market. It discovered two major basic strategic patterns, which are confirmed in a current study by Harvard University:

Strategy one can be characterized as market-driven internationalization: 'Go where the market is'. The sales and production departments of large firms have been operating according to this motto for decades. This striving for close spatial proximity to international markets and customers has been intensified in the 1990s. The industrial research units are adapting themselves to the global tendencies towards decentralization as well as on-site support of sales, production and development, the emphasis is on learning in 'lead markets' and adapting to local customer needs.

The second main reason for internationalizing research is to establish a presence in the 'hot spots' of world technological development. If one does not want to risk missing out on important knowledge obtained elsewhere in the world due to a narrow outlook, or unnecessarily repeat the effort to produce the results, one must go right to the 'hot spots' of technological activity. This alone opens up access to the resources, results and talents of the world's best research facilities.

Both basic strategic patterns follow the principle—if the relocation of research work promises advantages in the way of cost, without sacrifices in output and quality, it is an added motive for managerial decisions.

International locations of Daimler-Benz research

So much for the international state of the art. Now to the company:

The Research and Technology Division of Daimler-Benz AG has six sites in Germany. Since 1994 it has added three international sites to them. The Research and Technology Centre founded in Palo Alto in the state of California in the US in 1995 is primarily oriented to the second strategic goal: On-scene presence in the world's technological hubs. Its location close to Stanford University and Silicon valley enabled it to build a competence centre close to one of the most important hot spots in the field of information technology.

The centre has around 25 employees to do research there on transportation technology, information and communication techniques, and micro-technology. At the same time, the centre helps the company detect new technological and social trends and thus allow new products to be matched to customer and user needs more speedily and precisely.

A result of the alternative strategy—research where the market is—are the two research centres in Oregon, USA and in Shanghai. The research centre launched in Shanghai in 1995 is a laboratory run jointly with the Shanghai Institute of Metallurgy. The institute assists the microelectronics subsidiary Temic, which in the course of its internationalization has appreciably expanded its presence in China over the years by founding several joint ventures. The tasks centre around topics from the field of 'electronic packaging'. This is a decisive problem area in the further development of microelectronics, as several limitations like volume, weight, cost, and reliability are caused by the packaging.

In addition, a service centre was set up to support ongoing production processes at Temic departments, directly and at short notice through analyses and investigating faults. The integration of the Shanghai institute into our ongoing microelectronics research in Germany is ensured by the linking of themes and regular exchange of Chinese and German scientists.

Since the beginning of last year a 'Vehicle Systems Technology Centre' in Portland, Oregon was set up. It is located at the site of the commercial vehicles subsidiary, Freightliner, in order to support the activities of this company in the North American commercial vehicle market. The objective is to ensure that the specific conditions of the American market are taken into account in the early stages of research in product development.

Currently twelve employees—experts in mechanical science, electronics, information technology and control systems theory—support Freightliner in designing future electronic systems for vehicle diagnosis or driver assistance and adapting advanced tools and methods to purposes of vehicle development. A further focus is the human factors engineering, the assessment and optimization of the operation and control of technical systems with regard to the abilities and needs of humans—the customers. These are all topics to which research attaches increasingly great importance for vehicle development work of the future.

The Vehicle Systems Technology Centre relies in its tasks on close coordination with the vehicle research staff in Germany and with other worldwide research activities of the group. These three institutes alone, in the USA and China—none more than two years old—show how up-to-date and fresh its internationalization efforts are. Still younger, however, is the fourth research facility which was inaugurated in Bangalore in 1997—the Daimler-Benz Research Centre India. With this institute the company is simultaneously pursuing both strategies mentioned.

Firstly it sees India as a promising country—as one of the hot spots—for research in the area of information technology. In particular, the group wants to make use of the great know-how in communications technology, software engineering and mathematics. From this know-how an enormous impetus for fields of the future important to the company such as multi-media, pattern recognition, telematics and manufacturing solutions are expected.

An important component of the company's concept is a close working relationship with Indian universities and research establishments as well as with the software industry based in this country. For instance it hopes to:

- Involve scientists from the top Indian institutes in its research projects on a sabbatical basis.
- To place research contracts with high quality local partners.
- To give Indian students an opportunity to work on tasks of its research centre by commissioning degree dissertations and doctoral theses.

At the same time it wants to facilitate entry to new markets and build long-term relationships with the group subsidiaries in the region, thereby signalling a long-term partnership based commitment of the group to this fast advancing country.

Further measures towards internationalisation of Daimler-Benz research

But internationalization of research means much more to Daimler-Benz than just building research institutes abroad. To work in the vanguard of research it must simultaneously seek cooperation with a large number of international competence centres. This includes:

- Cooperation in projects with key universities worldwide
- Strategic alliances both with other enterprises and within international programmes
- Participation in high-tech venture capital funds
- Cooperation with suppliers of research services
- Participation in sponsorship programmes.

Daimler-Benz is active in all these areas.

But these are not the only ways to benefit from the knowledge generated in the world's 'hot spots'. I must also mention the technology field offices, the circle member group, and our international auditor network.

The idea to have technology field offices was born at the end of the 1980s when, in the course of diversification, the group branched out into new technologies, and an international technological monitoring was considered to be necessary. This led beginning in 1998, to the establishment of technology field offices in Japan, the CIS and the USA. They consist of small teams of three to five people whose mission can be described as:

- Technology monitoring, i.e. observation of core technology fields and identification of changing trends
- Preparing the ground for cooperation in research
- Building and maintaining networks of experts.

The offices also provide services on behalf of the group like preparing visits, procuring information or carrying out research.

For the systematic involvement of leading international experts it has in addition created the instrument of the circle member group. This is a circle of 25 currently top scientists from leading universities of the world, through whom in the medium term, the group aims to set up personal ties with important contacts. At the same time, the members of the circle provide a continuous overview of international progress in areas of research important to Daimler-Benz.

For the external evaluation of research topics and results, a network of research auditors has been established. It now comprises 55 of the world's leading scientists and, beyond the original intention, has also developed into a first rate group of consultants.

Challenges

I think the necessity of strategic internationalization of research, oriented to regional market needs and worldwide technological developments, is indisputable for a global player like Daimler-Benz which has developed a great deal of dynamism in this respect, the benefit of which will be considerable. However, the demands too, are enormous. Here I see two essential tasks:

First when geographical borders are bridged, cultural distance must be overcome simultaneously. Research facilities distributed across the world or international cooperation can produce a real added benefit only in a multicultural climate. That does not happen by itself but must be actively furthered. Daimler-Benz research therefore supports the described internationalization schemes with a bundle of personnel related measures, which include sending German researchers abroad for limited periods, increasing the percentage of foreign employees at the German locations and the systematic formation of scientific teams with an international composition.

The second challenge resulting from the increasing importance of worldwide alliances is the need to make use of the opportunities provided by information technology. This applies not only to collaboration in research, but extends much farther encompassing the entire product gestation process which is becoming more and more complex as product development takes place within an integrated worldwide system. Daimler-Benz[1] is currently carrying out a test project in which we are using various methods and systems of data, video and audio communication with a view to allowing parallel development of a vehicle at several locations.

The challenges of globalized research are great but there can be no mistake about it, the route will continue to be followed the world over. I would even say that in future the internationalization of industrial research will also have to be understood as internationalization of the structures of cooperation. Companies can profit from this frontierless cooperation across the entire spectrum—not just in the field of research. And above all, this collaboration will foster mutual understanding between different cultures, to everyone's benefit.

[1]Daimler-Benz merged in 1998 with Chrysler and is now called DaimlerChrysler.

Future Trends in the Exploration, Use and Misuse of Outer Space

U.R. RAO

Developments in space

As one enters the third millennium, one may look back at the amazing discoveries and developments in the field of space science and technology during the last four decades, since the birth of the space age initiated by the launch of the Russian Sputnik in 1957. The planet earth is now often called the global village, because as one views the earth from the vastness of space, it is truly a small village. While space has no doubt established human connectivity across the entire planet extending even to the remotest corners of the world, the real question is, what is the nature of this global village?

The rapid developments in space technology, spurred initially by the severe competition between the two super powers provided a great impetus to space science resulting in spectacular discoveries. Space technology opened up the entire window of the electromagnetic spectrum to the scientists who were till then essentially viewing the cosmos through the intervening atmosphere primarily in the narrow optical window. This enabled space scientists to discover entirely new celestial objects in outer space which were predominantly emitting radiation in the form of ultraviolet, infrared, X-ray and gamma rays. Exotic celestial objects like pulsars, neutron stars, quasars and even black holes, which

*Sixth Bose-Einstein Lecture, 26 February 1998, India International Centre, New Delhi.

were not even dreamt of, were discovered in quick succession taking advantage of the availability of space platforms. Scientists could trace back the history of the universe all the way to the so called 'Big Bang' beginning through the faint microwave remnant from the birth of the universe and extend their vision to the end of the universe. Today one is excited about the possibility of life on Mars. A few planetary systems, other than our own, have been discovered adding to the speculation on the existence of intelligent life elsewhere in the universe. Planets and planetary atmospheres, compact objects and galaxies, stars as they are born and those which are exploding have now become accessible to scientists. While one did not even know that the earth had a magnetosphere, scientists are now able to look at magnetospheres of solar system planets and distant stars.

It was Ludwig Biermann of Germany who originally hypothesized the existence of a continuously blowing supersonic plasma stream, later named as solar wind by E. Parker, by examining the dynamics of comet tails. The successful discovery of the continuously blowing solar wind oozing out of the sun at a tremendous velocity of 300 to 1,000 km per second by Mariner-II in 1966, with which I was also associated, provided the first proof of its existence. Very soon it was clear that this tenuous plasma stream is what provides the energetics for many of the spectacular phenomena happening on the earth such as the magnificent aurora, magnetic storms, electrical breakdowns, etc. It is now known that Venus is so hot that even lead can melt. Even more, with the sulphuric acid raining all the time, it is such a hell that I do not think anyone would like to go there. Scientists have been able to virtually stare at Comet Halley from the Giotto spacecraft and look at its physical and chemical composition. They have already begun discussion on the Kuiper Express which will take a close look at the asteroid belt. No doubt in the not distant future, one will be able to realize the dream of Tsilkovasky, who prophesized that 'One day we shall be able to ride on the asteroids'. Space technology has opened up the entire universe providing, for the first time, a proper perspective of the significance of human beings on this planet. Hopefully, mankind will not only discover life but even intelligence elsewhere in the universe in the coming decades.

Even more important is the ability of space platforms to take a synoptic look at vast areas on the earth which has enabled researchers to utilize space technology capabilities for communication, navigation, education, meteorological forecasting, disaster management, management of natural resources and environmental monitoring. Successful application of space remote sensing for looking at land use, agricultural monitoring, ocean dynamics, mineral prospecting, urbanization and environmental degradation have enabled space scientists to initiate sustainable integrated development at micro level. Continuous monitoring of the increase in green house gases, ozone depletion, global warming, coastal area degradation and deforestation have provided the quantitative estimate of the alarming manipulation of the environment by anthropogenic activities. The global conference on environment held in 1992 at Rio de Janeiro was to focus primarily on the environmental aspects of anthropogenic activities, which, if left unchecked, could result in the destabilization of the green house equilibrium with catastrophic consequences for humanity as a whole.

Remote sensing and meteorological satellites have become invaluable for providing timely information on natural disasters to enable governments to take appropriate steps in terms of disaster management as also evolving strategies, involving structural changes, for long term mitigation of such disasters. Spectacular progress has been achieved in remote sensing technology which is already providing repetitive imageries at 5 m resolution and is expected to provide 1 m resolution imageries in the next couple of years. Ocean monitoring satellites are providing information on the dynamics, temperature distribution, surface winds, moisture content and phyto-plankton density across the oceans. These inputs derived from using optical as well as active microwave remote sensing techniques are not only providing vital inputs for meteorological forecasting, but also for identifying areas of fish rich shoals to maximize fish catch. The data from the Indian remote sensing satellites IRS-1C and IRS-1D, which are undoubtedly the best remote sensing satellites in the contemporary civilian world today, are now regularly being received in Germany, through a cooperative arrangement. This is the second such ground station to be established in a foreign country, after the EOSAT station in Norman, Oklahoma in the USA.

Rapid development in silicon technology, computers, power sources and inertial systems have resulted in substantially reducing the weight of satellites. Mini-satellites of 500 kg class, 100 kg class of micro-satellites and even 10 kg nano-satellites have already become operational to carry out limited application missions in a cost effective way. Space scientists now talk of femto satellites of a few grams to micrograms which could be launched in tens of thousands, as a swarm of mosquitoes, their output could be combined using well known aperture synthesis techniques, thus adding to the romance of space exploration.

Space technology has virtually revolutionized global communication by initiating the so-called third wave information revolution. Information revolution, in turn, has been responsible for the emergence of a demassified society which is primarily based on specialized production centres, segmented niche markets and customized services as against mass production, mass manufacture, mass distribution, mass entertainment and mass education, characteristic of the second wave industrial revolution. With information being generated at 200 trillion bits a day, one necessarily needs information on information or so-called secondary information. Information and databases have become the modern encyclopaedia. Development of powerful global as well as national information infrastructures, which can be quickly accessed at will anywhere on land, air, or sea has become imperative for all nations and industrial enterprises wishing to successfully compete in the knowledge based global marketplace.

The satellite communication era began in 1965 with the launching of the Early Bird, the first communication satellite, which had just about 240 voice channels. Spectacular developments in digital compression techniques and satellite technology have made it possible to provide satellites with a capacity of 20,000 voice channels in addition to the TV channels. Thanks to digital compression techniques, it is now possible to transmit video at 4 mega bits/second and CD quality audio at 32 kilo bits per second. Developments in terrestrial communication systems have resulted in local area networking which combined with satellite communication have become the backbone of multimedia services across a wide area and metropolitan area networking systems with satellite based VSATs providing the connectivity. While Internet communication provides the link to the outside world, Intranet and Extranet systems have paved the way for virtual private networking. Global mobile personal communication systems with low earth, intermediate and even geostationary satellites are being introduced. The distinction between audio, video, data and message has completely vanished due to the confluence of powerful computers and communication technology. Video on demand, DTH, bandwidth on demand have become the buzz words of modern communication society.

The satellite communication market is expected to grow from the present base of US $ 80 billion a year to over US $ 300 billion a year by 2010. The greatest expansion necessarily has to be in the developing countries because of their poor infrastructure. In India alone, the number of telephones, for example, which will be added in the next 10 years to increase the telephone density from the present 1.4 per 100 to say 8 or 10 per 100, will be more than the number of telephones in all the developed countries put together. The growth rate of VSAT networking, which has been doubling every year, is a clear example of the expansion which is likely to happen in developing countries. No wonder the global market is focusing its market expansion particularly on India and China, the two large developing countries which are poised on the threshold of a communication explosion.

The overcrowding of the geostationary arc at 36,000 km above the equator, where all geostationary communication satellites are located, was considered to be a limiting factor for a long time. Technology solutions such as frequency reuse, use of multiple polarizations, collocations of satellites, shift to higher frequencies, etc., have greatly eased the problem. The introduction of digital compression techniques have made it possible to use one transponder for broadcast services where ten were needed earlier, thus considerably lessening the demand as well as cost of the space segment. Global and regional personal mobile communication systems, in combination with cellular systems, turned the dream of providing human connectivity anywhere across the world on land, air or sea into a reality. Digital audio broadcasting has already become a reality over Africa with the operationalization of AfriStar and within the next year is expected to become available over the Asian and American continents.

With the Global Positioning System (GPS) based navigation systems steadily gaining in importance over the last two decades, the concept of free flight, which has been the goal of every aeronautical engineer, is poised to take off. The free flight concept is based on the use of the GPS system, for navigation along with collision avoidance technology and the geographical information system which allows the flight commander to choose his own flight path without requiring air traffic control. With the avoidance of human intervention, which is prone to error, and the pilot's ability to choose the optimal path, free flight could not only be expected to be safer but also result in at least 25 per cent saving in aviation fuel.

The state of the global village

It is now widely recognized that the second wave industrial revolution was essentially based on maximizing the output without caring for the conservation of natural resources and hence was basically unsustainable. If only one had carefully computed the economic benefits taking into account the externalization factors, one would have recognized that historically humankind has traded its collective human labour against a much larger entitlement of commodity bundles, the negative balance being accounted against over-exploitation of natural resources. The green revolution has no doubt helped in increasing the global food grain output from about 600 million tons in 1950 to almost 2,000 million tons in 1995. Only one third of this increase is due to an increase in the area of cultivation, the rest being attributable to increased irrigation, use of hybrid cultivars and application of chemical fertilizers and pesticides. However, the negative repercussions of the very green revolution due to extensive water logging, inadequate drainage and indiscriminate use of chemicals and fertilizers have resulted in serious degradation of over 1,200 million ha of land globally. Widespread deforestation to the tune of almost 17 million ha per year has caused severe soil erosion, resulting in increased run-off of rain precipitation, land degradation and lack of adequate water recharging all over the world.

Over 100 million ha of prime agricultural land in India today is degraded. The once fertile agricultural belt in the Indo-Gangetic plain has become highly saline. As a result 60 per cent of the sugarcane growing areas in Uttar Pradesh and Bihar produce only 30 tons/ha, just one-third of that produced in Karnataka, Maharashtra and Andhra Pradesh, which yield on an average 90 tons/ha. Water management particularly in the developing countries is so bad that nearly 25 per cent of the people do not have access to clean drinking water. Every day 25,000 people die due to water borne diseases and almost one-fourth of the global population is susceptible to malarial infection. Increased soil erosion from 1 ton/ha to almost 10 tons/ha due to severe deforestation has made the Brahmaputra the largest sediment carrying river in the world.

While the availability of additional land for agriculture has dwindled to less than 0.2 per cent per year, the potential for increased irrigation is limited. The population in developing countries continues to grow exponentially. Even the most optimistic estimates indicate that the global population is likely to reach 11 billion by 2050 of which over 9.5 billion will be residing in the developing countries. The explosive growth in population is primarily confined to developing nations which are already suffering from large scale illiteracy, poor agricultural productivity, industrial backwardness and poor quality of life. The population in India is expected to cross 1.6 billion by the middle of the next century. Even at the present level per capita food intake of just 2,400 calories per day, which is much less than what is recommended by the FAO, the developing countries are forced to import over 120 million tons of food grains. Even assuming the growth in food grain production continues at 2 per cent per year, which is a very optimistic assumption, the developing countries would need to import over 400 million tons of food grains annually by 2050. India will have to import 30-50 million tons of food grains per year by 2020, even assuming the growth of food production continues at 2.7 per cent per year.

While space has shrunk in time and distance transforming the planet into a global village, one will have to inquire into the real nature of this global village. It is certainly not a homogeneous happy village, where everybody on the planet is assured of a minimal decent quality of life. 80 per cent of the global population residing in the developing countries contribute less than 15 per cent of the global GDP, with just 20 per cent of the people residing in the industrialized world accounting for over 85 per cent of the global GDP. India is a typical case, while having 16 per cent of the global population, she possesses only 2 per cent of the area, accounts for less than 3 per cent of the global energy production and contributes just 1.2 per cent of the global GDP. The annual per capita income in India is US $ 400, less than a fortieth of that in the USA. The amount of per capita arable land available in India is just around 0.15 ha as against over 1 ha in developed countries. Of the Indian population 45 per cent continue to be illiterate and 25 per cent do not even have access to clean drinking water. The agricultural productivity in the developing nations varies between 0.5 tons to 2.5 tons per ha as against 4.5 tons per ha in industrialized nations. Considering the explosive population growth in the developing nations the greatest challenge in the coming decades is to provide adequate food, health, economic and environmental security to all the people in the world. The question is: Can space technology provide solutions to evolve a truly equitable global village?

Space for sustainable development

In the modern era where science and technology have become the most powerful currency of power, and the ability to instantaneously access the vast information and databases has become vital for competing in the globalized market place, it is only the optimal use of the developments of space communication that can provide the comparative advantage to a nation in achieving economic security. With the phenomenal developments in the growth of satellite communication, India today has the necessary infrastructure to establish the required human connectivity across the entire nation. The poor export market of just around US $ 40 billion a year and the low productivity and efficiency of its industrial infrastructure, rated on a low 39 among 80 industrial nations, is partly due to the earlier emphasis on inward looking import substitution policies, but mostly it is due to large scale illiteracy. The liberalization policy adopted in 1991, if sincerely followed, will help India in global competition and will also release a large part of public funds for tackling societal problems such as education, sanitation, environment, and health care through privatization of transport, communication and energy infrastructure, on which so far almost 50 per cent of plan outlay has been expended.

Eradication of large scale illiteracy, particularly female illiteracy, is now universally recognized as the key to development. Accumulated evidence clearly indicates that family planning can only be successful by improving female literacy. Starting with the SITE programme in 1975, India has made considerable progress in the use of satellite technology for providing distance based developmental education to remote areas. Recent experiments conducted on a small scale have established the feasibility of one way video and two way audio distance education centres for providing local and culture specific, conventional, continuing, as well as special education even to the population in remote areas. A fairly large scale experiment to disseminate a wide variety of edu-cational programmes on health, hygiene, family planning, environment, and better agricultural methods to the predominantly tribal population in the Jhabua district of Madhya Pradesh is now being successfully carried out. The answer to the massive problem of eradication of illiteracy truly lies in the large scale implementation of distance education using dedicated satellites.

The only way to ensure food security for the people is by substantially increasing the productivity on a sustainable basis. The food grain productivity in India is on an average around 1.6 to 1.8 tons per ha, much less than productivity in industrialized countries, which have achieved a figure of over 4.5 tons per ha. Calculations indicate that the productivity figure in India and other developing countries has to reach a minimum of 4.5 tons per ha by 2050 if large scale starvation is to be avoided. Such a development has to be completely sustainable which means it must be accompanied by adequate measures for conservation of soil, conservation of water and respect for environmental integrity. The initiation of sustainable integrated development requires the combined use of biotechnological and space inputs. Vital information is provided by space remote sensing on soil moisture, soil type, land use practices, agriculture, vegetation, command area irrigation, and waste land configuration. Flood and draught disasters need to be carefully analysed for each watershed to evolve strategies for conservation of soil and water.

Combining space inputs with biotechnological inputs involving dwarf varieties of hybrid cultivars, genetic seeds, integrated pest and fertilizer management strategies and appropriate fertigation practices it is possible to achieve enhanced productivity of food grains. Continuous monitoring from space will provide advance information on the onset of any degradation enabling scientists to take appropriate mid-course correction, thus ensuring the long term sustainability of agricultural practices. Actual field experience gained from the initiation of sustainable integrated development strategy over a number of watersheds in India has clearly demonstrated that it is indeed possible to double the food grain production over the next 20 years by combining biotechnological inputs with information derived from space remote sensing.

Misuse of space—international mechanisms for preventing misuse

Phenomenal developments in space technology have fully established their capability to improve the quality of life of people by providing the most effective means for communication, broadcasting, navigation, eradication of illiteracy, improved weather prediction, disaster monitoring, management of natural resources and initiation of sustainable integrated development. However, like any other sophisticated technology, space technology is also prone to massive misuse. The establishment of the UN Committee on Peaceful Uses of Outer Space (UN-COPUOS) in 1959, which has 61 member states, was essentially meant to regulate the use of space for peaceful purposes and ensure that all nations benefit from the use of this technology. UN-COPUOS took the first significant step in the development of space law in 1967, with the adoption of the 'Treaty on Principles Governing the Activities of States in the Exploration and Use of Outer Space, Including the Moon and other Celestial Bodies' which is often referred to as the Magna Carta of space. Since then, the UN-COPUOS assisted by its two sub-committees, the scientific and the legal, has adopted five multilateral treaties which include agreements on the rescue of astronauts, international liability for damage caused by space objects, registration of objects launched into outer space and the agreement governing activities of states on the moon and other celestial bodies. In addition to the five treaties, four sets of legal principles governing activities of states in the exploration and use of outer space, the use of artificial satellites for international direct television broadcasting, the principles relating to the remote sensing of the earth from space and the use of nuclear power source in outer space, have also been agreed upon. Responding to the needs of the countries and realities of the day, UN-COPUOS was successfully democratized two years ago, with the chairmanship of the UN-COPUOS going to a developing country for the first time in its history, and this author was unanimously elected as the chairman of the UN-COPUOS in 1997.

Even though the membership of the UN-COPUOS is at present limited to 61 countries, two UNISPACE conferences involving all UN member states were held under its overall direction, the first in 1968 with Dr. Vikram Sarabhai as the scientific chairman of UNISPACE-I and the second in 1982 with Prof. Yash Pal as the general secretary. These two conferences were instrumental in bringing awareness of the potential of space technology applications to the entire world and in actively promoting international cooperation to enable the vast benefits of space technology applications to reach all countries, developing countries in particular. The fall of the Berlin Wall, the end of the cold war and increasing commercialization of space technology, including the launch scenario, have opened up new opportunities for all countries to benefit from space. Meanwhile the number of countries utilizing space technology have also increased substantially, even though the degree of utilization and capability differs widely from country to country. The changed international order notwithstanding, UN-COPUOS strongly felt the need to organize a third UNISPACE conference open to participation of all member states of the world in July 1999, under this author's chairmanship.

The Outer Space Treaty defines a launching state as a state providing for the launch vehicle, the state whose spacecraft is launched or the state from whose territory the launching takes place. With commercial entities taking over the manufacture of spacecrafts and launching them and the scene of launching shifting to platforms on open seas, need has arisen to redefine the outer space legal framework. While the Space Charter states that space shall be used for peaceful purposes and for the benefit of all nations, the definition of peaceful purposes continues to be a matter of interpretation. Fortunately, the Charter forbids the placing of nuclear weapons and weapons of mass destruction in space. Unlike the moon and the celestial bodies, which have been reserved by common consent for 'exclusively' peaceful purposes, the word 'exclusively' is deleted as far as the earth's orbit is concerned. This has automatically legitimized spying as a necessary means for keeping watch on non-peaceful activities. Rapid developments in technology now allows one to virtually look into somebody else's backyard through electronic, optical, micro-wave and infrared eyes. The use of knowledge derived from remotely sensed information on mineral wealth, agricultural production or economic activities in other countries for private profit or exploitation has made a number of developing countries, which do not possess adequate expertise, very vulnerable. The misuse of direct broadcasting from space for a cultural invasion of traditional societies is no more a fiction.

The appropriation of the geostationary arc at 36,000 km above the equator, from where all geostationary communication satellites operate, has been and continues to be on the principle of 'first come, first served'. The developing countries, who were late in taking advantage of the benefits of space technology, today find themselves in a precarious position, unable to get a slot allotted for operating their own satellites. A number of equatorial countries on the other hand have put in a territorial claim for the geostationary arc above their country, based on arguments of sovereignty. The demarcation between air space or the delimitation of space still remains an open question. In the meanwhile a few nations have either already obtained or filed for a number of slots in the geostationary arc which they may not be able to use. A notable example is Tonga which has filed for 27 slots. Unethical, real estate selling of geostationary slots has already begun. I am afraid many of these issues will continue to be debated in the years to come and new issues will be added to the list. The brighter side of the picture is that a mechanism exists where all multilateral issues can be debated and a reasonable solution could hopefully be found to ensure peaceful development of space activities.

Whereas the Convention of International Liability for Damage Caused by Space Objects of 1972 is reasonably clear regarding the liability for damage to property and people on the earth, the liability for damage to space objects, which has now become a distinct possibility with numerous objects flying in space, is at best nebulous. The space debris problem has indeed become a serious hazard with estimated millions of debris dispersed in space of which more than 40,000 objects, larger than 10 cm, have actually been tracked. Any collision of these objects moving with a velocity of something like 8 km per second with an active space satellite, can be truly catastrophic. UN-COPUOS has finally started discussion on space debris and it is hoped that a satisfactory solution will be found to minimize the debris hazard at least in the future.

The Moon Treaty (Agreement Governing the Activities of States on the Moon and Other Celestial Bodies, 1979) specifically states that these celestial bodies, which form the common heritage of humankind, can only be used for exclusively peaceful purposes and for the benefit of the entire human race. When this treaty was adopted, commercial exploitation of the moon and other celestial bodies was not considered a practical feasibility. The picture has significantly changed with the development of technology and it is indeed within the realm of reality to think of commercial exploitation of these celestial bodies in the years to come. While the cost of launching into the geostationary orbit is about US $ 30, 000 per kg with present day rocket technology, newer technologies being developed aim at reducing the cost by an order of magnitude to about US $ 3,000 or less per kg. Already a number of travel agencies are busy selling intercontinental travel tickets for US $ 10,000. The reduction of launch costs is the key factor which will facilitate the commercial exploitation of the moon and other celestial bodies or even the establishment of space habitats, and large space manufacturing facilities. The question is, when will commercial exploitation of the moon become a reality and how does one ensure that the benefits from such commercialization will reach the entire humankind? The question of intellectual property rights, laws applicable on the moon, maintenance of law and order and a host of other issues are yet to be agreed upon. The process of law making which has generally been a post-facto reaction response needs to become proactive, anticipating developments which are within the realm of technological reality.

The principles relating to remote sensing of the earth from space, while allowing unrestricted remote sensing, demands that the sensed state must have the priority for obtaining the data on its territory. There is, however, no mechanism to ensure this because under the pretext of cloud cover or some other technical reason the sensing state can deny the imagery. Since remote sensing has essentially been used for meeting societal requirements, it has not yet become fully commercial. Remote sensing activities, so far, have been subsidized by the states which have operated such satellites. The scenario, however, is fast changing, with the entry of a number of commercial operators building satellites which can image at 1 m resolution. The commercial operators expect to reap substantial gains by selling value added services in addition to imageries to a vast customer base which is likely to include insurance companies, real estate developers, planners, disaster management outfits, etc. Today over 800 reconnaissance satellites have been launched primarily by the USA and the former USSR. With improved capability, the dividing line between the traditional spy satellites and the civilian satellites has significantly narrowed to such an extent that serious doubts have been expressed regarding the need to build and operate expensive spy satellites. Commercialization of high resolution imaging, which will make them readily available to anybody willing to pay the price, is bound to create new security concerns across the world.

The Space Charter of 1967, which deals with the principles of the exploration and use of outer space for peaceful purposes, does not explicitly prohibit non-peaceful activities. The interpretation of this treaty by the super powers does not prohibit military activities, as long as they do not constitute an aggressive posture. With a pile up of over 20,000 nuclear warheads by the USA and Russia and a lesser number by Great Britain, France and China and the advances made in ballistic missile technology, space has indeed become a major area of threat to peace in the world. Even though the end of the cold war and SALT treaties have tried to lessen the threat, and the star war scenario having receded, attempts are being made again to renew the star war race. The principle of no first use of nuclear armament has not been accepted by all the nuclear powers. The list of nuclear powers is growing with Israel, India, Pakistan and other countries ready to join the club. The Committee on Disarmament with 61 member states, which was specifically created in 1979 with the mandate to deal with nuclear threats, ban on production of fissile material, nuclear disarmament, environmental modification, production and stockpiling of chemical weapons, and prevention of the arms race in space has failed to make any headway, except in the case of chemical weapons. Anti-ballistic missiles are being stockpiled by the missile powers. The Committee on Disarmament continues its endless debates on these issues, forming committees and sub-committees in a typical bureaucratic fashion, without being able to halt the arms race in space. The Chemical Weapons Treaty has been signed by

160 countries, but only 47 have ratified it so far. The tragedy is that the US and Russia, which have the largest stockpiles, are yet to ratify the treaty. While the romance of the space age has been truly exciting, the misuse of space for destructive purposes threatens the very existence of humankind on this planet. Continuous vigil is required to ensure that the vast potential of space is used only for achieving sustainable growth and providing food, health, economic and environmental security to all people and is not misused for destructive purposes.

I hope UNISPACE-III will deal with these issues and also provide the necessary impetus to enable all countries, developing countries in particular, to maximally benefit from the use of space technology. As India has consistently argued over the last three decades, the benefit from the use of such high technology can truly accrue to any nation only when its indigenous capability is established and a degree of self-reliance is built. It is extremely important to channelize international cooperation to promote the application of high technology for national development on a self-reliant basis. The best hope lies in making UNISPACE-III, in which all member nations of the world will participate, succeed in realizing its stated objectives for achieving a better and safer future.

Plate 13: Uwe Thomas, State Secretary in the German Ministry of Education and Research, explains German science policy for the twenty-first century—10th Bose-Einstein Lecture.
Photo: S.K.Sharma

Plate 14: S.N. Bose, drawing by an unknown artist.

Plate 15: Albert Einstein, a drawing by Frank Dyson.

Plate 16: R.A. Mashelkar, Director General of CSIR, speaks to a captive audience on science and technology and their impact on political and economic power, 11th Bose-Einstein Lecture.
Photo: S.K.Sharma

Comets: Fascinating Cosmic Objects

REIMAR LÜST

Both, Bose and Einstein were outstanding physicists. Albert Einstein is regarded as the greatest scientist of our century. The famous French physicist de Broglie wrote in 1925: 'The scientific world hung on every one of Einstein's words, for he was then at the peak of his fame'.

In the summer of 1924, he received a short paper from Satyendra Nath Bose, at that time a physicist of Dhaka University on 'Planck's Law and the Hypothesis of Light Quanta', which considered radiation as a form of a gas consisting of photons. Einstein was so impressed with the paper that he himself translated it into German and sent it to the editor of the *Zeitschrift für Physik* who published it in September 1924. At that time German was the language of the physicists. The reason for Einstein's interest was simple. He had seen immediately that it was possible to extend Bose's statistical methods to ordinary atoms—'Bose-Einstein Statistics' as they became known.

Bose was born on 1 January 1894 in Calcutta. He graduated from the University of Calcutta, taught there in 1916 and at the University of Dhaka from 1921–45 and again in Calcutta 1945–56. He died on 4 February 1974.

The Bose-Einstein Statistics accounts for instance for the frictionless creeping of superfluid helium. Just last year, the physicists have been fascinated by new experiments where the 'Bose-Einstein condensation' of sodium atoms could be observed. This phenomenon occurs at extremely low

*Ninth Bose-Einstein Lecture, 17 November 1998, India International Centre, New Delhi.

temperature (2μK), that means a temperature very near to the absolute zero point minus 273 degree Celsius.

The phenomenon which I will discuss today is also occurring in a space, where the temperatures are rather low, but certainly not as low as just mentioned. It is the space where our earth is moving around the sun, the so-called interplanetary space.

I do not know how many people have consciously seen the comet Hale-Bopp last year. Very rarely in the past was such a bright comet visible in the evening sky in such a good position for a period of several weeks. Even if only a small proportion of the public was aware of being so privileged, they took part in this astronomical event—something that has not occurred for decades.

The nature of comets

What kind of bodies are comets and to which parts of the observed astronomical families do they belong? Until today about 1,000 different comets have been observed, and every year 5 to 10 new comets are detected.

It is not surprising therefore, that throughout history the unpredictable appearance of comets has been puzzling mankind and created fear in the hearts of observers. A comet was regarded as an omen presaging disaster. Many people shared Aristotle's view that the appearance of a comet signaled adversity or drought.

Reports on the appearances of comets reach back many centuries before Christ. These early records come almost exclusively from China, Japan and Korea. A Babylonian description, interpreted as a reference to the comet of 1140 BC has also been found.

The question was whether comets were celestial objects or phenomena of the atmosphere. The Babylonians as also the Greeks at first were of the opinion that comets were cosmic bodies like planets. But these ideas changed with Aristotle who ruled out the planetary nature of comets. According to his theory, planets belong to the sublunar sphere, being the product of meteorological processes in our atmosphere.

Around the time of Christ's birth, the Roman Seneca, Emperor Nero's teacher, wrote: 'Some day there will arise a man who will demonstrate in what regions of the heavens comets take their way; why they journey so far apart from other planets, what is their size, their nature.'

Around 1600 years later, the time was ripe for scholars to work out the comets' rightful place on the sky. Tycho Brahe, generally considered to have been the greatest astronomer of his days, made the decisive observation by determining the parallax of a comet in 1577. He showed that the distance of this comet was considerably greater than the distance from the earth to the moon. Although he was not able to determine the orbit of the comet, these objects could no longer be placed in the sublunar atmospheric space. Since that time comets have been acknowledged as real objects in the solar system. But still there remained the question of their orbit. Although Kepler was so successful in determining the orbits of the planets, he was wrong in his belief that comets moved along straight lines, but with irregular speed.

Isaac Newton and Edmund Halley, the latter having been eclipsed in history by the more famous Newton, made the final step in describing and calculating the orbits of comets. It was Halley who encouraged the retiring Newton to publish the principles of gravitation he had developed after years of thought, supposedly inspired by the legendary falling apple. Newton recognized that gravity on earth represented the same law of force as that affecting the motion of planets around the sun. Halley edited the manuscript and financed the publication of Newton's great book, *The Mathematical Principles of Natural Philosophy* in 1687. But in applying Newton's laws of gravitation in correctly predicting the return of a comet, Halley went an important step further. It turned out to be the first direct confirmation of Newton's theories.

Halley compared the orbits of 24 comets on the basis of observations. He was able to show that these bodies move in elliptical orbits like the planets, though some were very elongated, hardly distinguishable from parabolas. From his comparisons he detected that the orbit of the comet in 1682 was almost identical with those of earlier years (1607 and 1531). Convinced that these three visits had been made by the same comet, he predicted that he would return in 1759. He knew that he would not live to see if his calculations proved correct. But he expressed the hope that 'candid posterity will not refuse to acknowledge that this was first discovered by an Englishman'. Indeed on Christmas night in 1759, almost 17 years after Halley's death, the comet was seen again just as Halley had predicted. This comet now carries his name.

No other comet has been observed over such a long period. From the time this comet was mentioned for the first time—by the Chinese in the year 240 BC—it has been seen at each periodic visit. The missing record of an observation in the year 164 BC has most likely been found just recently on a Babylonian table with cuneiform characters. In 12 BC Halley's comet appeared over Rome and was said to have presaged the death of Agrippa. In AD 66 it was seen over the city of Jerusalem before the city was destroyed. It appeared again in AD 451 during the battle of Chalons, when the Roman General Aetius defeated Atilla the Hun, and it appeared in AD 1066 during the battle of Hastings and was blamed for the defeat of King Harold's armies by William the Conqueror. When it returned in 1910, comet Halley caused considerable public concern—especially when it was known that the earth was actually to pass through the comet's tail. Some fearful people went to great lengths to avoid being harmed by the poisonous cyanogen gas of which the tail was composed.

The English astronomer Halley had been able to show that comets move around the sun in elliptical orbits and therefore belong to our planetary system. However, until 35 years ago it was still an open question where to place comets in the solar system and how they originated. A remarkable fact was that the cometary orbits were not in a plane around the sun, like planetary orbits but were distributed in a sphere around the sun.

The orbits of comets and their origin

In 1950 the Dutch astronomer Jan Oort analysed 19 cometary orbits, which had very elongated ellipses and orbital periods of more than 30,000 years. The cometary orbits were very accurately determined. Oort's analysis showed that the majority of these comets come from regions situated in the outermost part of our solar system. These regions extend from about 30,000 AU to about 50,000 AU (50,000 AU corresponds to 0.8 light year—the distance to the second star, Proxima Centauri, is 4.3 light years). Oort concluded from his analysis of planetary orbits that a cloud of comets must exist in this region.

In the meantime, the use of electronic computers had made it possible to analyse a much greater number of cometary orbits, as well as to determine their orbits with greater accuracy. These calculations have confirmed the ideas of Oort. The cloud contains roughly 1,012 comets with a total mass of 1,028 g or about the mass of planet earth.

This comet cloud came into being, together with the planets and our own sun, some four and a half thousand million years ago, through contraction and condensation from an interstellar cloud. A great number of these comets have since then been far away from the sun, and because of the very low temperatures out there, have remained practically unchanged as if they were in an astronomical deep freeze. Consequently, the material in a comet could be representative of the very earliest condensations in our solar system, and observations of its physical and chemical properties could yield information about the nature of our solar system in its early stage.

But how does a comet from this cloud come close to the sun so that it can be observed from the earth? The most likely explanation is that a star streaking past the solar system from time to time throws comets in this cloud out of their usual circular orbit into an elliptical one. This elliptical orbit could then be affected so strongly by the large planets that a comet comes so close enough to the sun, to become finally visible. It can be estimated that about once every two million years other stars come close enough to our solar system for their gravitational force to have a detectable effect on the comet cloud.

The physical and chemical properties of comets

So what are comets? The observer on earth sees a brilliantly shining object, which as it gets closer to the sun shows the tail that is characteristic of comets. It was supposed that the brilliance comes from a temporary atmosphere surrounding a solid nucleus. From the earth it was impossible to observe this nucleus directly, and one only had clues; however it has only been possible for the last few years to receive radar signals from comets suggesting that they had a radius of about 1 km.

From the observations of the outer envelope, the American astronomer Fred Whipple derived the theory that the core was a dirty snowball, i.e. a conglomerate of ice and dirt. With a density of about 1 g/m^3 and a diameter of a few kilometres a core like this would have a mass of 1,015 to 1,018 g, so that the total mass of all 1,012 comets comes roughly to the order of magnitude of only the earth's mass.

If one of these nuclei—of the kind postulated by Fred Whipple—approaches the sun, gas is given off by its surface as it heats up. This gas pulls off particles of dust, and that builds up a constantly expanding atmosphere, a coma, which can reach dimensions of 10,000 to 100,000 km. The ratio of gas to dust varies from one comet to another and depends on the distance from the sun; it can range from ten times more gas to equal proportions. Gas emission begins to develop at about the distance of Jupiter (5 AU) with the lighter gases such as carbon monoxide. At the distance of Mars water vapour can also be observed.

As the comet gets closer to the sun a tail develops always pointing away from the sun. Until the early 1950s there was no theoretical understanding of the observations of comet tails. The German astrophysicist Ludwig Biermann found the answer to this puzzle in the early 1950s. It was a striking fact that many comets build up not one but two tails. One tail consists of particles of dust that are accelerated by the pressure of sunlight and hence blown away in the opposite direction from the sun. The question was how the very small gas particles accelerated and how the very long tails of gas which sometimes stretched for more than 100 million km, come about? It was Biermann who from the existence of this electrically charged tail of gas and high velocities noted in it (more than 100 km/sec), concluded that this phenomenon could be caused only by a corpuscular radiation from the sun. Ludwig Biermann thus developed the theory that a continuous stream of particles must be emitted from the sun like a wind. This stream of particles was indeed found, in the early 1960s, by the first interplanetary space probe, and since then it has been known as the solar wind.

The solar wind is responsible for the comet's ionized tail, and the force it exerts on the gas particles is far stronger than the radiation pressure on the dust particles. Because of this the dust tails are curved and do not have velocities found in the almost straight, long tails of gas.

Artificial cometary tails—experiments in space

In 1957 and 1958 the Russians and the Americans launched the first artificial satellites. Very soon space probes were also launched equipped with measuring instruments to probe the space where planets and comets were moving. As Biermann had predicted the existence of solar wind could be proven and its strength could be measured for the first time. But how did solar wind interact with the cometary tails? So far it was not possible to observe this phenomenon directly. Also at that time it did not seem feasible to send a space probe to a comet and carry out measurements in its neighbourhood in particular within a cometary tail.

For this reason the Max-Planck-Institut for Extraterrestrial Physics developed the idea to create an artificial cometary tail in space in order to study its interaction with the solar wind.

It would, of course, have been ideal if it had been possible to inject into space the same atoms or molecules that had been found in the comet tails. First estimates showed however, that several tons of carbon monoxide would be needed to produce in space a cloud of CO+ visible from the earth. Since there was no hope of finding a rocket to carry such a heavy payload, this idea had to be abandoned. On the other hand, the element barium was very well suited to the purpose: A few grams of barium would be enough to make a visible cloud, it would be rapidly ionized by sunlight, and the barium ions could—when illuminated by the sun—be observed from the earth.

Since at that time only small, high altitude sounding rockets were available it was decided to use the upper atmosphere for this experiment at a height of 200 to 250 km. After various unsuccessful experiments in 1962 and 1963, the first barium cloud was released into the ionosphere above the Algerian Sahara in November 1964.

Though the interest in comet physics was the starting point for the barium cloud experiments this new method has been and is still used for quite different scientific purposes by numerous groups in other countries, for studying the earth's upper atmosphere. By using barium clouds it is possible to visualize the lines of force in the earth's magnetic field, just as one makes lines of magnetic force visible here on earth by using iron filings. The observation of barium clouds allowed for the first time reliable measurements of electrical fields and of features of the plasma high up in the earth's atmosphere. Outside the earth's atmosphere, but still within its magnetosphere, only two barium cloud experiments were carried out in the 1960s, using ESA satellite HEOS and a big scout rocket made available by NASA. But it was still not possible to reach inter-planetary space and consequently the solar wind.

Finally, Gerhard Haerendel and his group at the Max-Planck-Institut at Garching were able to achieve the original aim of the barium cloud technique. On 27 December 1984 and 18 July 1985, two barium clouds were created in the solar wind outside the earth's magnetosphere at a distance of 17.2 RE and 18.6 R~.

The opportunity for these experiments presented itself during the implementation of the Active Magnetospheric Particle Tracer Explorer (AMPTE) mission, a joint project of the USA, West Germany and the United Kingdom. Although the prime aim of this mission was the long-range tracing of the transport of ions from the solar wind into the magnetosphere and further redistribution therein, it was ideally suited for the study of the initial interaction of the injected plasma with the ambient medium. To this end a set of plasma and field diagnostic instruments were installed on the German Ion Release Module (IRM) which carried barium and lithium release containers. Furthermore, a subsatellite developed in the United Kingdom (UKS) was injected into nearly the same orbit as the IRM in order to provide another probe for the plasma and field perturbations created during the release experiments. In addition to the *in-situ* diagnostics, optical observations from the earth contributed valuable information on the dynamics of the solar wind interaction with the seeded plasma. Each time 2 kg of barium were released and it was possible to observe the interaction of the artificial cloud with the solar wind and the surrounding magnetic field. Now these measurements can be compared with the observations in the near neighbourhood of comet Halley from the spacecrafts sent to this comet.

The preparation of the space missions to Halley's comet and the spacecrafts

Four space agencies—the Intercosmos of the USSR Academy of Sciences, the Japanese Institute of Space and Astronautical

Science (ISAS), the National Aeronautics and Space Administration (NASA) and the European Space Agency (ESA)—sent spacecrafts to Halley's comet or have been involved with Halley observations from space during the comet's last apparition in 1986. Intercosmos launched Vega-1 and Vega-2, ISAS launched Sakigake and Suisei, the ESA launched Giotto, while NASA was able to redirect a spacecraft which was formerly known as the International Sun-Earth Explorer ISEE-3 and was renamed Inter-national Cometary Explorer, ICE.

The spacecrafts complemented each other in flyby distance ranging from 600 to 7 million km and comet heliocentric distances at the time of encounter ranging from 0.79 to 0.89 AU. The encounters all took place in March 1986, because Halley crossed the ecliptic plane on 10 March and the launch energy required was minimized to enable the encounter spacecraft to stay close to the ecliptic. The relative flyby speeds, i.e. the speeds in the comets frame of reference, were all very high. Unfortunately, due to Halley's retrograde orbit, only fast flyby missions were possible.

The programmes of the different space agencies were coordinated by an Inter-Agency Consultative Group (IACG) which was formed in 1981 and has since then met very regularly.

Comet Halley was chosen for a number of good reasons:

1. Compared to most of the other comets, the data of its orbit were well known.
2. The energy requirements for the spacecraft to reach comet Halley were relatively low.
3. Although comet Halley has often approached the sun, its dust and gas production rate can be compared with those of 'new' comets.
4. Last but not the least, its name and the role it played in history was attractive.

The Italian painter Giotto di Bondone saw Halley in 1301 and was so impressed by its appearance that he incorporated it a few years later as the 'Star of Bethlehem' in one of the frescoes in the Scrovegni Chapel in Padua. The painting shows details of the comet's coma and the tail not unlike the drawings made by scientists in the nineteenth century. This is why ESA baptized its spacecraft 'Giotto'.

The scientific objectives for the mission to comet Halley were the following:

1. To measure the production of the dust and the gases, the size and the shape of the dust particles as well as to define the chemical and isotopic mixing of the gas and the dust particles.
2. To investigate the physical processes and chemical reactions in the surroundings of the comet as well as the interaction of the comet with the solar wind plasma.
3. To discover the nucleus and if it really exists to take photos of it with a resolution of upto 50 m.

The scientific experiments on the various spacecrafts complemented and supported each other.

The Giotto spacecraft was spin-stabilized, with a nominal spin period of 4 sec. Its diameter was 1.86 m, and the height from the tip of the tripod to the bottom adapter ring was 2.85 m. At launch Giotto weighed 960 kg.

The Giotto mission

After five years' preparatory work and a great many tests, Giotto was launched on schedule from Kourou in French Guiana on 2 July 1985 by the European Ariane rocket. Its flight was to last nine months. Giotto was put into its orbit so accurately that minor course corrections were only needed a couple of days before the flyby. During its flight, all instruments were switched on and tested several times.

The nine months of the flight were full of tension for everyone concerned, and were not entirely free of worries. The least worry was that Halley's comet would suddenly disappear. But

I shall not easily forget the early morning of 24 January when I was woken up with the news that radio contact to Giotto had been lost. As it happened to be exactly the date on which the American Voyager probe was to fly by the planet Uranus, my team did not for a while have immediate access to the big American reflector telescopes for sending radio commands to Giotto. At 2 o'clock the following morning I was relieved to hear from California that radio contact with Giotto had been re-established.

At long last everything was ready for the moment when Giotto was to fly by the core of Halley's comet, at 3 minutes past midnight on 14 March. It was planned that Giotto should fly by at a distance of 500 km, plus 40 km, on the sun side of comet's nucleus. Final corrections were to be made on 9, 11 and 12 March. To do this, however exact data on the comet's orbit were needed; but the normal astronomical observations made on the ground were inadequate. Therefore the work had been shared internationally over a long period between Intercosmos and NASA as agreed in the Inter-Agency Consultative Group.

The two Russian probes Vega 1 and 2, which flew by Halley on 6 and 9 March at a distance of 8,890 to 8,030 km, pinpointed the exact position of the comet, and the data were then passed from the Russian control centre in Moscow to the European control centre in Darmstadt. However, one also needed to know the exact position of the two Russian probes, which could be determined only with the help of the big American antennas. In this way Halley's position was plotted within 40 km. It showed that after almost 9 months flight and travelling 4.4 million km Giotto would, if no course corrections were made from the ground, pass the comet at a distance of 800 km, plus or minus 20 km.

Among the experimenters, opinions differed as to how near Giotto should come to the comet, and they argued about the closest distance until 24 hours before the event. The camera team, wanted to get no closer than 1,000 km, while most of the other experimenters wanted a distance of only 500 km.

They were prepared to take the risk of the probe being totally destroyed by the impact of dust particles, despite its protective shield. In the end all concerned agreed on a distance of 540 km. Ground control consequently made a slight course correction, 24 hours before the planned time of flyby. The probe was by then 140 million km away from the earth and a radio signal needed about 8 minutes to travel this distance.

During the night of 13 to 14 March 1986 this stage was reached, after the first comet particles had already been recorded on 12 March at a distance of rather more than 7 million km from the core. The experimenters were able to follow the instrument data being received at Darmstadt in real time and these were released to the public a few minutes later. All the experiments worked without a hitch. The distance from the comet's core actually was 605 km.

Though Giotto was meant to flyby very close to the comet's core, it was expected that it would survive. Between 8 and 16 seconds before the closest encounter Giotto was struck by a relatively large dust particle, which made the probe rock. Because of this the antenna was no longer pointing exactly at the earth, and radio contact was suddenly lost. Such an eventuality had however been foreseen by the engineers, as they had built nutation dampers into Giotto, 32 minutes later these dampers had reduced the spacecraft's nutating movements sufficiently for the variation in antenna pointing to be once more within 1 degree of the required direction, and radio contact was restored.

The two Japanese spacecrafts had a flyby distance of 1,51,000 km on 8 March and of 7 million km on 11 March, while the American Explorer ICE had a distance of about 28 million km on 25 March.

A decisive element for the success of the space mission was the camera, developed by the Max-Planck-Institut für Aeronomie in Germany. The quality of the camera proved to be even better than expected. The passing velocity of Giotto with respect to the nucleus of comet Halley was 2,50,000 km/h. The camera had to take pictures from such a fast moving object. A comparison would demonstrate the performance of the camera: One would be able to photograph a pilot in an airplane passing at a distance of 160 m with sound velocity or equal to 1,200 km/h with a resolution of 4 mm.

For the first time the nucleus of a comet could be photographed. Its surface as well as its size and shape could be observed and measured. As the theory predicted, it is comparable to a dirty snowball and it is one of the darkest bodies in the solar system. The pictures show a lengthy not spherical nucleus, comparable to a potato. The long axis has a length of about 15 km, while the smaller one between 7 to 10 km. The surface is irregular and shows spherical structures, quite comparable to craters, valleys and hills. Besides these pictures, the other measuring instruments like mass spectrographs and dust counters have given very important results about the nature of this comet.

Giotto suffered some damages during the flyby due to the impact of dust particles. After the encounter the instruments were checked, all worked perfectly with the exception of the camera, which apparently was damaged by the dust impact. It was possible to retarget Giotto to return to the neighbourhood of the earth in July 1990. Using an earth gravity assist it could be redirected towards another comet, namely comet Grigg-Skjellerup. On 10 June 1992 Giotto passed this second comet at a distance of about 200 km and again interesting and important results were received on the earth. At this time Giotto was at a distance of 314 million km from the earth and had travelled about 64 billion km. But Giotto still had enough fuel and its orbit has been redirected again to come close to the earth in the year 1999. Then it will be woken up from its hibernation, the instruments will be checked again and Giotto might be redirected to another comet.

But let me return to Hale-Bopp: Observations of this comet have given new results as well. But while comet Halley will return after 76 years to the vicinity of the sun and hopefully will be observed again, this encounter will take much longer for comet Hale-Bopp. When Hale-Bopp passed the point closest to the sun, it lost about 100 tons of material every second. Yet even after this heavy shedding it has sufficient material to show up again at least a dozen times. But its next visit will occur in 2380 years, that means at the end of the forty-fourth century.

Is Delhi Dying? A Pragmatic Approach for a Clean and Healthy Delhi

KIRAN BEDI

Delhi has been one of the great centres of human civilization, famous for its beauty, architecture and culture. Today Delhi is in danger.

The area of contemporary Delhi in the triangle between the river Yamuna and the northern and southern ridge, on the transit routes from north to south and east to west, saw at least seven cities—or fifteen, or more? It is assumed that Indraprastha, the legendary capital of the Pandavas, scene of the epic Mahabharata, is buried under the Purana Qila, the 'Old Fort'. Dynasties rose and perished and each founded a new capital, from Lal Kot near the Qutub Minar (about 1052) to Siri, the capital of the Khiljis (about 1302), the first Tughlaq ruler Ghiyas-ud-din moved south and built Tughlaqabad, his son Mohammed Tughlaq founded Jahanpanah before he unsuccessfully tried to transfer the capital to central India. His successor, Feroze Shah, moved north again to what is today Feroze Shah Kotla, then on the banks of the river and the Moghul Emperor Humayun built his capital Dinpanah with the Purana Qila in the sixteenth century on the soil of Indraprastha. These frequent changes don't necessarily reflect pride and whims of kings, but they were probably caused by changes in the course of the river, failed monsoons and scarcity of water, problems which haunt Delhi even today. The seventh city Shahjahanabad, built by Emperor Shah Jahan

*Fourth Bose-Einstein Lecture, 12 September 1997, India International Centre, New Delhi.

from 1638 onwards, was the most splendid capital, vibrating with the arts and sciences, a brief happy period, when Shah Jahan had a Persian couplet engraved in his audience hall: 'If there is a paradise on earth, it is here, here and here'. The British destroyed much of Shahjahanabad, but added New Delhi,[1] the so-called Lutyen's Delhi, which served as the seat of government for British-India since 1932. Delhi in its entirety is now the capital of independent India and with its many historic places, its wide avenues, parks and recreational areas it is truly the most beautiful metropolis in modern Asia.

However, Delhi has during the last 50 years grown beyond its capacity (Figure 1) and when a city becomes a metropolis, the human dimension is lost. This means critical changes in complex human relationships, which tend to become impersonal, apathetic, selfish and disinterested.

The capital of independent India has grown from 1.744 million inhabitants in 1951 to 12 million at the turn of the century and may well be home to 14 million people by 2001. Every year 5,00,000 people are added, 2,00,000 by natural growth while the rest through migration. Alongside its political and administrative importance the capital has also attracted the arts and science, and commerce and industry. But the fast growth is mainly a fall out of underdevelopment in some of the neighbouring states. People come in search of work and livelihood (Table 1).

Needless to say, Delhi is bursting at the seams and civic amenities are insufficient—there is shortage of power and water, lack of an efficient public transport system, while hospitals, schools, and housing cannot keep pace. Hence the amorphous growth of slums and an alarming increase of begging and of crime, heightened by the multiplying of monkeys and street dogs, which harrass human beings. Pollution has reached intolerable levels—Delhi is now the fourth most polluted city in the world and may before long become the most polluted place worldwide. 1,999 tons of

[1] Edwin Lutyens and Herbert Baker were the architects of New Delhi.

TABLE 1: THE GROWING CONCERNS OF DELHI

Issue	*1991*	*1997*	*2001*
Population	94.2 lakhs	115 lakhs	140 lakhs
Pollution (ranking in the world)	4	2	1
Vehicles	19 lakhs	26.8 lakhs	40 lakhs
Shortage of drinking water (daily)	2.2 crore gallons	20 crore gallons	22 crore gallons
Shortage of electricity	105 MW	350 MW	408 MW
Shortage of hospital beds	7,300	22.930	33,350
Unemployment	9 lakhs	11 lakhs	34 lakhs
Population living in slums	13 lakhs	22 lakhs	48 lakhs
Average speed of vehicles	45 kmph	30 kmph	16 kmph

Note: 1 lakh = 1,00,000; 1 crore = 10 million.

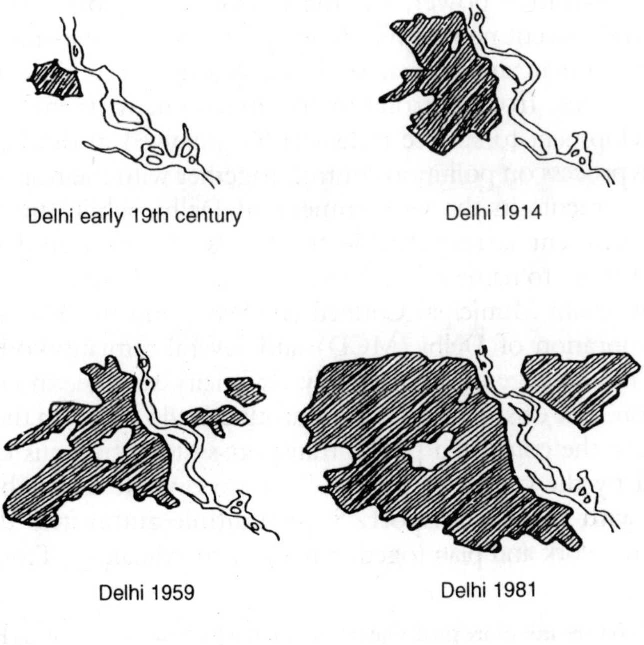

Figure 1: Growth of Delhi.

pollutants are emitted every day, compared to 100 tons a decade ago.

The scarcities of daily life led to a phenomenon half-jokingly called 'internal secession', meaning, whoever can afford it, buys a private vehicle, operates a generator, digs his/her own well and uses courier services instead of the post office. As such every year pressure on existing facilities gets even more intensified.

Shortsightedness and lethargy on the part of the decision makers could well be one of the reasons for the present dilemma. There are however structural factors, which may partly be responsible. Delhi is not a normal state like West Bengal or Tamil Nadu. Since 1993 it is a National Territory, an improvement on its previous status of UnionTerritory.[2] Delhi now has a legislative assembly and a state government. But being the national capital, the centre has reserved some administrative power, i.e. the police is controlled by the central government, while railways, telecommunication and environment are central subjects. No less than six central ministries, from petroleum to environment, from urban development to surface transport are involved in the legislation process on pollution control, together with the respective departments in the Government of Delhi, while the state government is responsible for roads, drainage and land aquisition, to name a few. There are even two local bodies, the New Delhi Municipal Council (NDMC) and the Municipal Corporation of Delhi (MCD) and several semiautonomous bodies incharge of water supply, electricity, sewage, and sanitation. The existing ring railway around Delhi could, in theory, relieve the congested public transport system, but it is operated by the central ministry of railways, while feeder buses are with Delhi Transport. Such multiple authorities don't always work and plan together in close coordination (Table 2).

[2] Union Territories are small units which do not have the same political representation as the states. They are administered by the union government through a Lt. Governor, i.e. Chandigarh or Diu (editor's note).

TABLE 2: FUNCTIONAL RESPONSIBILITIES OF DIFFERENT GOVERNMENT ORGANIZATIONS FOR PREVENTION AND CONTROL OF POLLUTION IN DELHI

Type of pollution	*Nodal Ministry/Govt. organization*
1. Vehicular pollution	— Ministry of Petroleum and Natural Gas (introduction of cleaner fuels, supply of 2T-oils, fuel adulteration, unleaded petrol, low sulphur fuel, CNG/LPG, etc.
	— Ministry of Surface Transport (legislation on motor vehicles, vehicular emission norms, penalties, etc.)
	— Ministry of Urban Development (mass rapid transport system, urban planning and development)
	— Ministry of Industry (industrial policy and promotion, industrial development)
	— Ministry of Home Affairs (for Govt. of NCT of Delhi) (traffic Police)
	— Ministry of Environment & Forests (MOEF) (protection of environment; recommendation regarding emission standards, surveys on air quality monitoring, etc.)
	— Govt. of Delhi (DPCC, State Department of Environment, Transport Department/Public Works Department)
2. Industrial pollution	— Delhi Pollution Control Committee (DPCC), State Department of Environment, Ministry of Industry, MOEF (standards, surveys, monitoring, etc.)
3. Domestic waste water and municipal solid wastes	— MCD/NDMC, Ministry of Urban Development, State Department of Environment, DPCC, Water Supply and Sewerage Undertaking of Govt. of Delhi, MOEF, etc.
4. Industrial solid wastes	— State Department of Environment, DPCC, Industrial Department, Ministry of Power, DESU, NTPC, MOEF, etc.
5. Noise pollution	— State Department of Environment, DPCC, Transport Department of Delhi, Delhi Police, MOEF, etc.

The city's main problems are: water, energy, air, transport, and waste.

1. WATER: The supply of water falls short by 2,00,000 gallons daily. Water runs from the household taps only for a few hours a day, at the peak of summer it often fails

completely. The main source of water is the Yamuna river, which is tapped at Wazirabad before it enters Delhi and channelled into a treatment plant. It does provide reasonably clean water. In middle-class homes tap-water for drinking will be boiled and/or run through filtering equipment, the more affluent families buy mineral water drawn from distant sources in the mountains or even from Pondicherry more than 2,000 km away! Needless to say, the slum dwellers often don't have access to water taps at all. The river runs 48 km through Delhi and becomes a slow drain for nine months of the year. 264 million gallons of industrial sewage are added every day, the city's open drains release 500 million gallons of waste water and sewage into the Yamuna.

2. ENERGY: Delhi is short of electricity to the tune of 350 MW per day. There are four reasons for this shortage: *a*. Delhi does not generate enough power and is therefore dependent on importing power from other states, notably from Haryana; *b*. Much of the shortage, however, is home-made in Delhi itself. Illegal tapping of power leads to losses of 40-50 per cent and not only poor people pilfer electricity, industries and even the owners of posh farmhouses in the south of Delhi are equally culpable; *c*. In addition, maintenance is poor and equipment often substandard; *d*. Last not least faulty meter reading adds to the losses. Consecutively the citizens suffer frequent power breaks.

3. AIR: Delhi is now considered to be the fourth most polluted city in the world. About 2,000 metric tons of pollutants are emitted every day.

The contributors are:

Vehicles	64 per cent	Industry	12 per cent
Thermal Plants	15 per cent	Domestic	7 per cent

The dust concentration is unusually high compared to other cities in the world, depending on the location, the season and the time of the day.

Suspended matter microgrammes/cubicmetre:

Mumbai	228	Delhi	700
Kanpur	488	London	221
Calcutta	527	New York	178

Two million vehicles are plying on Delhi's roads and 2,00,000 are added every year, not counting vehicles in transit from other states. The main polluters are buses, trucks and auto-rickshaws with their two-stroke engines. Poor maintenane of vehicles and adulteration of petrol or diesel are the major reasons for the intense pollution.

4. WASTE: Delhi produces close to 6,000 metric tons of solid waste every day.

On an average each resident produces 0.4 kg of garbage every day, 12 kg per month, or 144 kg annually. That is actually much less than what residents of highly industrialized countries produce in the way of garbage, but still, Delhi's three garbage dumps are insufficient. Again garbage is not being collected regularly, as a result wild garbage dumps appear all over the place.

Action to be taken

Neither are the citizens helpless nor are civil servants without powers.

The need of the hour is to create awareness, citizens and administrators alike must learn to think environmentally, in a holistic and futuristic way.

A 'Delhi Renewal Plan' is going to be launched soon together with a 'Delhi Renewal Task Force'. The first thrust will be on the reduction of vehicular emissions. The old buses of the Delhi Transport Corporation must be replaced by buses with an improved design. Similarly vehicles older than 15 years need to be phased out. *Action*: Delhi Government.

Stricter vehicular emission norms corresponding to Euro-II will have to be enforced effective from year 2000.

Action: Ministry of Surface Transport and Ministry of Industries.

Cleaner fuels need to be introduced like propane, CNG, alcohol and battery operated motors. *Action*: Ministry of Petroleum and Natural Gas, and Delhi Government.

The checking of adulteration of fuel is an urgent task. *Action*: Ministry of Petroleum.

Furthermore the implementation of regular and rigorous inspection and maintenance programmes for all types of vehicles has to be made compulsory. *Action*: Transport Authority.

As can be seen, several central ministries have to work in close coordination with each other along with the Delhi Government.

Some steps have already been taken, i.e. pollution control of private vehicles in a three month cycle are pescribed by law. Only leadfree petrol is being sold at Delhi's petrol pumps.

However as yet only the symptoms are being cured, the main task is to improve the public transport system. The existing circular railway could be effectively used. A high capacity bus system would relieve some of the congestion. A mass rapid transport system has to be built and construction has actually started.

Regarding water and the renewal of the Yamuna river several actions are necessary:

Firstly sewage needs to be pumped to the full capacity of the existing treatment plants and the sewers and pumping stations need to be regularly maintained. The entire urban settlements have still to be provided with sewers and sewage works. *Action*: Delhi Water Supply and Sewage Disposal Undertaking (DWSSDU). Furthermore hazardous waste (like used batteries) needs to be identified and disposal facilities have to be developed. *Action*: Delhi Pollution Control Committee (DPCC) and Delhi Government.

Some achievements in controlling the pollution of the Yamuna's 48 km Delhi stretch should however be noted:

14 sewage treatment plants are under renovation/upgradation/construction.

15 common effluent treatment plants in 20 industrial estates have been initiated for construction. All large and medium industries and some units of small scale industries have provided for effluent treatment plants.

The shifting of polluting industries from non-conforming to conforming areas is in progress.

Waste needs to be turned into wealth. The art of recycling is fairly advanced in India, but it still needs to be organized in a systematic manner. In order to do so the local authorities must take up door-to-door collection. A beginning has already been made in some areas, for which new jobs have been created with residents contributing to the cost. The collection and transportation system needs to be modernized, which is the responsibility of the MCD and the NDMC. Like in some industrialized countries of Europe it is necessary to create public awareness to the segregation of waste into biodegradable and solid waste, with the former recycled as fertilizer. Glass and paper should be collected separately as it is raw material. NGOs and local residents associations need to be involved to spread public awareness. Biomedical waste is still a great problem, however guidelines for the treatment and disposal of such waste including standards for incinerators have been developed. Rules for biomedical waste need to be notified. *Action*: Ministry of Environment and Forests. These rules must be enforced by local authorities and small hospitals and nursing homes have to set up common treatment facilities—a task for the Delhi Pollution Control Committee (DPCC). Following a public interest litigation the Supreme Court has directed the Central Pollution Control Board (CPCB) to monitor collection, transportation and disposal practices adopted by the MCD and the NDMC. What has been achieved is: The existing waste receptacles have been improved and are being better maintained than before, while the frequency of garbage lifting has been increased. The landfill sites have been improved. Awareness among the authorities concerned is being created.

Last not least, awareness of saving water needs to be spread. In spite of acute shortage it is still being wasted mindlessly.

NGOs can help better inform the citizens (Table 3). The media too should intensify their efforts to create environmental consciousness.

Noise pollution is a new concept for India, but with the availability of various technical devices noise levels have gone up dramatically in the cities. The standards and guidelines for ambient and vehicular noise, for noise control of appliances and public address systems as well as for diesel generator sets have been developed. But these standards need to be enforced. The use of loudspeakers in residential areas must be restricted

TABLE 3: UNINTENTIONAL WASTAGE OF WATER

Activity	*Method adopted*	*Qty. used Ltr.*	*Method to be adopted*	*Qty. required Ltr.*	*Qty. saved Ltr.*
Brushing Teeth	Running tap for 5 min.	45	Tumbler or Glass	0.5	44.5
Washing Hand	Running tap for 2 min.	18	Halffilled wash basin	2	16
Shaving	Running tap for 2 min.	18	Shaving mug	0.05	17.95
Shower	Letting shower run while soaping, staying under shower too long	90	Wet down, tap off, soap up, rinse off	20	70
Flushing	Using old fashioned large capacity cistern	13.5 or more	Dual system: short flush liquid waste, full flush solid waste	4.5 9	4.5 or more
Watering Plants	Running hose for 5 min.	120	Water can	3	115
Washing Floor	Running hose for 5 min.	200	Mop and bucket	18	182
Washing Car	Running hose for 10 min.	400	Buckets (two)	18	382

PLEASE USE WATER WITH CARE!

SAVE WATER WATER IS LIFE

Source: Panchtatva Garima Foundation—NGO for Environment, New Delhi.

and the use of stationary diesel generators prohibited. Unfortunately the public address systems of temples and mosques seem to compete with each other in decibels in the name of misunderstood traditions. Similarly political parties don't spare the eardrums of the voters.

There are numerous other issues, the one of utmost concern is the growth of slums following migration of labourers from other states. Builders are the greatest polluters. They recruit labour from Bihar, U.P., etc., but don't provide them with accommodation or sanitary facilities. Nobody builds for the poor. Hence they construct their own temporary huts, so-called jhuggis, in places where there is no water, electricity, or sanitation. By now about one-third of the city's population are slum dwellers. Every removal to a proper building costs the Delhi Government Rs. 50,000, obviously resources for such rehabilitation are scarce.

The 'Delhi Renewal Plan' alluded to earlier seeks to involve the citizens through residents' welfare bodies, womens' associations, transporters' unions, etc. More power should be given to the citizen wardens, whose number must grow from the 300 at present. The least citizens could do is to protect the existing trees and plant new environment friendly ones which improve the quality of air like *Banyan, Peepal, Ashoka, Neem,* and *Amla*, and help improve the environment.

Delhi has been destroyed many times in the past, but always a new city emerged. It is time we think about a renewed Delhi.

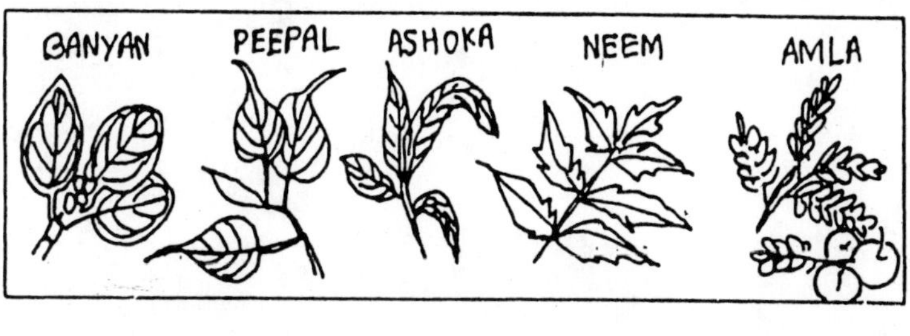

Figure 2: Panchavati.

Ecology and Economy: The German Experience

HEINRICH F. VON LERSNER

In Germany we experienced during the last ten years after unification a political and ecological experiment which is of exemplary value for both, ecology and economy. At the time of the German division the basic social and economic conditions of the two states were the same. Both were highly developed industrial countries with a population that had similar educational and social backgrounds. The standard of raw materials and agriculture was also on the same level.

Theoretically the policies for the protection of the environment should have been more advanced in East Germany, the former German Democratic Republic, than in West Germany, because awareness of the need for environmental protection had been raised there at an earlier stage. Moreover, this necessity could be supported by quotations from Karl Marx, Friedrich Engels and Vladimir I. Lenin. Protection of the soil had been included in the East German Constitution 26 years earlier than in the Federal Republic of Germany. An environmental ministry had been founded in East Germany 13 years before one was established in West Germany with the tolerance values in some areas being more stringent compared to those applied in the latter. In an authoritarian system enforcement of rules is easier, therefore East Germany should have been environmentally more progressive. However, after unification it turned out that in most areas of environmental protection, the GDR had stagnated at a level prevalent in West

*Second Bose-Einstein Lecture, 3 March 1997, India International Centre, New Delhi.

Germany 20 years earlier, when awareness of modern environmental protection became politically relevant. In East Germany the state of public sewage systems and sewage treatment plants, industrial waste, air and waste water cleaning technology had become obsolete while agriculture lagged behind. Better conditions were found only in such areas which had something to do with raw material shortage and motor traffic. However household waste generated in East Germany was only one-third of that generated in West German households, while the percentage of packages collected and recycled was higher in East Germany. And the rail/road ratio was much more advantageous. In the areas of domestic refuse and motor traffic the eastern part of Germany meanwhile reached the bad level of West Germany. The reasons why environmental policy in East Germany lagged behind West Germany's by about 20 years are as follows:

— West Germany had a higher gross national product. A prosperous country can make higher investments in environmental protection.
— Insufficient transparency of ecological data in East Germany. At the beginning of the 1980s the GDR declared all environmental data to be a state secret and because of the lack of transparency of data, industry, trade and consumers were not able to choose between environmentally friendly and less friendly methods and products.

II

In the Federal Republic two processes determined its environmental policy in the past 27 years, the period when I was allowed to render my advise. The *political process* started at the grassroots, triggered by the very visible pollution of the rivers, the smog in the cities and the damage to the forests, the so-called 'Waldsterben'. The sources of the poisonous emissions were often hundreds of kilometres away from the points where the damage occurred. This public outcry eventually reached the political decision-makers at the state and central level and policy-making for the protection of the environment started around 1970. Today global environmental damages have caught the attention of the international community, e.g. the global warming, changes in the world climate, the ozone hole, the expansion of the deserts and water shortages in arid zones. These problems now form a part of the international development cooperation.

The second, the *economic process* mirrors the conflict between environmental protection and economy in a market system. At the beginning we at the Federal Agency for the Environment (FAE) established so-called 'end of pipe' regulations defining the admissible outputs of waste water, exhaust air and items allowed to be thrown into waste containers. It became the duty of the enterpreneurs to comply with these rules. Later regulations covering production and products were issued and the Packaging Ordinance adopted to determine the behaviour of trade and consumers. An advanced environmental policy necessarily has an illiberal tendency. A comparison of both the German states around 1990 had however shown that a liberal market economy was superior to a state managed economy in all respects,including enforcement of environmental policies. The tension between economics and ecology can only be solved if the actors in the market: producers, traders and consumers behave as environmentally responsible as possible, so that the government need not interfere in their freedom.

III

I would like to give some examples of such voluntary free market instruments of advanced environmental policy. An already classic example was the use of asbestos in cement. In the late 1970s the FAE came to the conclusion that asbestos fibres in the air could cause cancer. It therefore required the prohibition of the use of asbestos. The government had some difficulties with this in view of the fact that in the Single

European Market, one can enforce common European requirements for products only through the European Union. As a consequence the cement industry promised to voluntarily refrain from using asbestos with cement in construction engineering by 1990, and in civil engineering by 1993. Independent experts were commissioned to supervise the compliance of this commitment. Hence this sector had already fulfilled its obligations before European regulations on the subject were adopted. There were similar voluntary commitments in the production and use of fluoride hydrocarbons which, as one knows, destroy the global ozone layer. German industry had already undertaken to cease the production of such substances before the period defined by the Montreal Convention came into force. The third example of such voluntary measures designed to avoid governmental interference with the freedom of the market forces was packaging, which used to waste an enormous amount of paper, plastic, glass and other materials, and moreover increased household garbage. According to the Packaging Ordinance of 1991, industry and trade were obliged to take back packages. This met with stiff opposition from trade. However, incentives were offered. If industry and trade took back and reused such packages within a voluntarily organized system and complied with certain quotas (e.g. 60 per cent of glass packages by 1993, 80 per cent by 1995, and 38 per cent of paper by 1993 and 80 per cent by 1995), the government promised to refrain from supervising the commitment through the police force by spot checksat every small shop to see, whether the assistant took back a yoghurt cup or not. For the time being this arrangement seems to work, although there are still some problems involving the law of competition, because the recycling industry has a near monopoly regarding the substances to be recycled.

An important instrument for such free market voluntary measures for environmental protection is the so-called eco-audit which was introduced by the European Union in 1994. According to this arrangement enterpreneurs could commission independent auditors to control the environmental compatibility of their production and products. They had to submit reports on these audits which could also be used by them for publicity purposes. This instrument is not only advantageous for the market, but it can also have deregulating effects. The government need not control all particularities of production and sale of such companies which voluntarily undergo an audit.

IV

Another aspect of environmental policies is its increasing internationalization and globalization. World trade is becoming liberalized according to the conventions of Marrakesh, which concluded the so-called Uruguay Round. As a consequence the importance of the requirements of environmental protection will increase in the newly established World Trade Organization (WTO). In particular highly developed industrial states will press for action against so-called eco-dumping, in other words against offers of less environmentally compatible products at lower prices.

The precondition for stronger harmonization of the requirements for products and their sale are, in particular, common standards for the ecological evaluation. In this respect the International Standardization Organization (ISO) is of utmost importance also in relation to environmental policy. One knows that in environmental protection also every producer and government tends to use the more advantageous statistics or analyses in their evaluations. In particular for the comparison of different alternatives, the so-called balance sheets on ecology, one needs international standards, according to which neutral and independent institutions could establish, which alternative was ecologically more favourable, e.g. the returnable glass bottle or the recycled plastic bottle, mineral oil as motor fuel for diesel vehicles or vegetable oil produced by agriculture. Such international standards are important for world trade, because this very environmental protection may sometimes be used to create trade hurdles, motivated by the wishes of national economies.

As the statistical evidence of the impact of carbon dioxide and other injurious gases on the world climate increases, the bigger will become the role of carbon dioxide in trade policies as well. As this substance also serves as an indicator of the consumption of fossil fuels, which are not renewable energies, one's responsibility towards the future generations demands its economical use. Every litre of oil and every kg of coal which one burns will one day be missed by future generations.

The Climate Convention, the first phases of which were adopted in Berlin in 1995, will be defined precisely in the years to come. The industrial states, which have the largest energy demand per capita, will have to drastically reduce carbon dioxide emissions not only in their own countries (about 50 per cent by the year 2050), but they will also have to support the sustainable development in less developed countries where energy demand due to industrialization and rapid population growth is increasing. However, satisfying these energy requirements should not be at the expense of the climate and non-renewable raw materials. Particularly for these reasons the importance of energy saving technology and utilization of solar, plant and wind energy will increase in world trade.

The conditions of the so-called Joint Implementation have still to be agreed upon even after further negotiations at the Climate Convention in Berlin in 1997, which will now be continued in Kyoto. Annually 916 million tons of carbon dioxide are released into the atmosphere from the territory of Germany compared to 742 million tons from India (as of 1993), which means that Germany's per capita share is 10 times higher than that of India. Given these facts there is clear evidence that the industrialized countries have to give financial and technical assistance to the less developed countries to reduce their rising emission levels.

V

Another area of environmental policy where international collaborative actions are required is the protection of the seas and transboundary rivers. This is a challenge both for Germans and the Indians as their countries are situated on big rivers and oceans. During my last visit to India ten years ago I had an opportunity to learn something about the pollution problems of the river Ganga. The big rivers in my country as well have not yet regained the quality one wishs. Thanks to international cooperation good progress has been achieved for the river Rhine, which took about 20 years. The eastern rivers Elbe and Oder, however still require extensive anti-pollution measures.

An important factor in this development is the time frame. If a municipality or a company knows that the law will eventually insist on the construction, e.g. of a sewage plant or a waste water treatment plant and they will have to build the same, it would be prudent to calculate this in their technical and economic decision making at an early stage.

Public pressure on companies for the protection of the seas only becomes stronger when catastrophes occur and are covered by the media. Some years ago a large number of seals died in the North Sea and a famous actor was seen on TV holding a dying seal in his arms. Suddenly it became possible to pass legislation in parliament on measures for the protection of the North Sea for which till then the necessary majority could not be obtained. Public pressure helped.

VI

The more the process from the local to the global level strengthens, the more important becomes the need for cooperation and exchange of international information. One must be able to exchange both the positive and negative experiences of technical and administrative instruments. Databases have to be internationally available. The FAE agreed on the exchange of staff members with some foreign countries. The expert for exhaust gases of the American EPA worked at the German Authority for a certain time, while his German colleague was at the EPA in Washington. Such an exchange of experts is of importance for industry and municipal authorities as well. Data bases, e.g. on the legal regulations, research results and on ecological effects of chemicals are at the FAE's disposal. Most of these data are accessible to the public. Language is still a hindrance in this sector. With the help of the European Environmental Agency in Copenhagen one hopes to be able to soon make these databases available in English.

There are however still several areas where the progress of environmental policy and the liberalization of world trade require closer collaboration among states. An example is the international trade in wastes which actually or supposedly are to be recycled. Increasing prices for an environmentally compatible treatment and disposal of waste within the industrialized countries will lead to increased exports of waste for recycling. Criminal activities in this area are as frequent as in the arms or drugs trade. For this reason authorities and companies have to work together to pool their information about details of the waste recycling in the importing countries. Such a cooperation was also the subject of discussion at the Basle Convention on the Transboundary Movement of Waste in 1989.

VII

The economic process also includes the production of eco-friendly goods. For Germany environmentally compatible technology has high significance for it to retain its competitiveness in the international markets. With a share of environmental technology in world trade in 1990 of 21.4 per cent Germany was ahead of the United States (17 per cent) and Japan (13 per cent). Since 1993 the USA has been on the first place. For Germany with a relatively high wages structure, to retain this market advantage is of utmost importance since outdated technical standards can be offered by states with lower wages on more advantageous conditions. It will retain its position in the world markets only if it offers state-of-the-art technology of tomorrow. In other words, environmentally sound and in particular energy and raw material saving technology.

I have already mentioned the time factor for environmental protection in market economies. In the European languages ecology and economy are sister terms. They are both derived from the Greek word *oikos* which means household. Since the beginning of the discussion on modern environmental policies the reconciliation of ecology and economy has been a hotly debated subject. Nevertheless this aim remains an Utopia like the Marxist aim of satisfying all human requirements. Economy means the management of man with his available assets, whereas ecology means the concern about nature's household in which the humans are just a factor but the most disturbing one. Therefore these sister terms will never become completely reconciled. Nonetheless one has to try to achieve this Utopia as closely as possible.

One of the main reasons for the conflict between economy and ecology are different time frames. The time needed for the soil to regenerate after harvesting is often not congruent with the date when the farmer has to repay his loans. Again the periods in which energy saving investment, e.g. in a house or in a factory becomes profitable are often longer than those in which the investors normally calculate.

For the time being one has to make every effort to develop instruments to bring these timings in line. Fiscal law can help if it grants tax incentives for energy saving measures. For example Germany has enacted a Waste Water Charges Act according to which discharges of waste water into lakes, rivers or the sea are subject to a charge. It has been increased from an initial annual rate and is still increasing in stages. But this charge must be spent by the authorities on the construction of sewage treatment plants. Unfortunately the

FAE has not yet been able to implement its proposal to increase the mineral oil tax in ten annual stages so that the buyer of a car and the investor in an energy-saving plant can calculate. However, as a regular political practice the Finance Minister just takes sudden decisions to hike the mineral oil tax whenever the government needs more funds. But what one needs is calculating periods of ten years. Industry always protests against new environmental obligations, therefore it would be very useful for them if such calculating periods were established for medium terms.

This is true as well for international harmonization of ecological requirements. Even if the manufacturers in a country are still allowed to use chlorofluorocarbon (CFC), which poses a serious threat to the climate, they will have to be prepared to expect that using these substances will soon be prohibited internationally due to the pressure of other countries where a ban is already in force. Therefore alternatives in a time bound frame will have to be found to save on the cost of conversion measures. It is essential in this respect to know in good time when which requirements are to be expected. One of the main aspects of harmony between ecology and economy is the reliability of timely information concerning environmental policies in the medium term.

VIII

Another conflict between economy and ecology has its roots in the widely diverging interests of the national economy, the protection of the environment, manufacturers and consumers. An old vehicle with a life of 25 years (instead of average ten years in Germany at present) has economic advantages. From the point of view of the national economy the long life of motor vehicles would have positive effects for the repair shops, but negative effects for the automobile industry. Older vehicles usually consume more petrol and hence are less environment friendly—the consumer has to make a choice between these factors. Experience shows that environmental protection creates more jobs. However what is the advantage for a worker in a chemical plant, which has been closed for environmental reasons, if a new job is created in a sewage plant 100 km away from his home?

IX

In a market economy special importance is given to the behaviour of trade and consumers. Personal examples as also the information spread by NGOs in environmental protection influence consumer behaviour. Industry only produces what trade will purchase and trade can only sell what the consumer is willing to buy. To give an example: German consumers refused to buy Indian leather goods for some time because they rejected the chemicals used in the manufacturing process. The Indian leather industry had to make changes. Citizens have been motivated to behave more environmentally responsible by social examples. The British MP who went by bicycle to No. 10 Downing Street to receive his appointment as Finance Minister helped social acceptance of the environment friendly bicycle. This holds true also for the German industrialist who on weekends would buy his rolls and always brought back the bag from the previous weekend. Specially in arid countries the rich often set a *bad* example as they want to demonstrate their wealth by lavishly watering their lawns all the day long in brilliant sunshine. In Germany the fast car is regarded as a symbol of prosperity and freedom. For this reason it is the only country in the world which does not know a general speed limit on motorways. We will have to change our habits, because in a world in which the space for humans is constantly decreasing with resources getting scarcer only that economy will have a chance to survive in the fiercely competitive world market which in early stages has learnt to produce, sell and consume goods as environmentally compatible as possible.

APPENDIX

Plancks Gesetz und Lichtquantenhypothese.

Von **Bose** (Dacca-University, Indien).

(Eingegangen am 2. Juli 1924.)

Der Phasenraum eines Lichtquants in bezug auf ein gegebenes Volumen wird in „Zellen" von der Größe h^3 aufgeteilt. Die Zahl der möglichen Verteilungen der Lichtquanten einer makroskopisch definierten Strahlung unter diese Zellen liefert die Entropie und damit alle thermodynamischen Eigenschaften der Strahlung.

Plancks Formel für die Verteilung der Energie in der Strahlung des schwarzen Körpers bildet den Ausgangspunkt für die Quantentheorie, welche in den letzten 20 Jahren entwickelt worden ist und in allen Gebieten der Physik reiche Früchte getragen hat. Seit der Publikation im Jahre 1901 sind viele Arten der Ableitung dieses Gesetzes vorgeschlagen worden. Es ist anerkannt, daß die fundamentalen Voraussetzungen der Quantentheorie unvereinbar sind mit den Gesetzen der klassischen Elektrodynamik. Alle bisherigen Ableitungen machen Gebrauch von der Relation

$$\rho_\nu d\nu = \frac{8\pi\nu^2 d\nu}{c^3} E,$$

d. h. von der Relation zwischen der Strahlungsdichte und der mittleren Energie eines Oszillators, und sie machen Annahmen über die Zahl der Freiheitsgrade des Äthers, wie sie in obige Gleichung eingeht (erster Faktor der rechten Seite). Dieser Faktor konnte jedoch nur aus der klassischen Theorie hergeleitet werden. Dies ist der unbefriedigende Punkt in allen Ableitungen, und es kann nicht wundernehmen, daß immer wieder Anstrengungen gemacht werden, eine Ableitung zu geben, die von diesem logischen Fehler frei ist.

Eine bemerkenswert elegante Ableitung ist von Einstein angegeben worden. Dieser hat den logischen Mangel aller bisherigen Ableitungen erkannt und versucht, die Formel unabhängig von der klassischen Theorie zu deduzieren. Von sehr einfachen Annahmen über den Energieaustausch zwischen Molekülen und Strahlungsfeld ausgehend, findet er die Relation

$$\rho_\nu = \frac{\alpha_{mn}}{e^{\frac{\varepsilon_m - \varepsilon_n}{kT}} - 1}.$$

Indessen muß er, um diese Formel mit der Planckschen in Übereinstimmung zu bringen, von Wiens Verschiebungsgesetz und Bohrs Korrespondenzprinzip Gebrauch machen. Wiens Gesetz ist auf die klassische

Theorie gegründet, und das Korrespondenzprinzip nimmt an, daß die Quantentheorie mit der klassischen Theorie in gewissen Grenzfällen übereinstimme.

In allen Fällen scheinen mir die Ableitungen nicht genügend logisch gerechtfertigt. Dagegen scheint mir die Lichtquantenhypothese in Verbindung mit der statistischen Mechanik (wie sie durch Planck den Bedürfnissen der Quantentheorie angepaßt worden ist) für die Ableitung des Gesetzes unabhängig von der klassischen Theorie hinreichend zu sein. Im folgenden will ich die Methode kurz skizzieren.

Die Strahlung sei in das Volumen V eingeschlossen und ihre Gesamtenergie E sei gegeben. Es seien verschiedene Sorten von Quanten von der jeweiligen Anzahl N_s und Energie $h\nu_s$ vorhanden ($s = 0$ bis $s = \infty$). Die Totalenergie E ist dann

$$E = \sum_s N_s h\nu_s = V\int \varrho_\nu d\nu. \tag{1}$$

Die Lösung des Problems verlangt dann die Bestimmung der N_s, welche ϱ_ν bestimmen. Wenn wir die Wahrscheinlichkeit für jede durch beliebige N_s charakterisierte Verteilung angeben können, dann wird die Lösung durch die Bedingung bestimmt, daß diese Wahrscheinlichkeit bei Wahrung der Nebenbedingung (1) ein Maximum sein soll. Diese Wahrscheinlichkeit wollen wir nun aufsuchen.

Das Quantum hat ein Moment vom Betrage $\frac{h\nu_s}{c}$ in der Richtung seiner Fortbewegung. Der Momentanzustand des Quantums wird charakterisiert durch seine Koordinaten x, y, z und die zugehörigen Momente $p_x p_y p_z$; diese sechs Größen können als Punktkoordinaten in einem sechsdimensionalen Raum aufgefaßt werden, wobei wir die Beziehung haben

$$p_x^2 + p_y^2 + p_z^2 = \frac{h^2\nu^2}{c^2},$$

vermöge welcher der genannte Punkt auf einer durch die Frequenz des Quants bestimmten Zylinderfläche zu bleiben gezwungen ist. Zum Frequenzbereich $d\nu_s$ gehört in diesem Sinne der Phasenraum

$$\int dx\, dy\, dz\, dp_x\, dp_y\, dp_z = V \cdot 4\pi\left(\frac{h\nu}{c}\right)^2 \frac{h\, d\nu}{c} = 4\pi \frac{h^3\nu^2}{c^3} V d\nu.$$

Wenn wir das gesamte Phasenvolumen in Zellen von der Größe h^3 einteilen, gehören zum Frequenzbereich $d\nu$ also $4\pi V \frac{\nu^2}{c^3} d\nu$ Zellen. In bezug auf die Art dieser Einteilung kann nichts Bestimmtes gesagt werden.

Indessen muß die Gesamtzahl der Zellen als die Zahl der möglichen Anordnungen eines Quants in dem gegebenen Volumen angesehen werden. Um der Tatsache der Polarisation Rechnung zu tragen, erscheint es dagegen geboten, diese Zahl noch mit 2 zu multiplizieren, so daß wir für die Zahl der zu $d\nu$ gehörigen Zellen $8\pi V \frac{\nu^2 d\nu}{c^3}$ erhalten.

Nun ist es einfach, die thermodynamische Wahrscheinlichkeit eines (makroskopisch definierten) Zustandes zu berechnen. Es sei N^s die Zahl der zum Frequenzbereich $d\nu^s$ gehörigen Quanten. Auf wie viele Arten können diese auf die zu $d\nu^s$ gehörigen Zellen verteilt werden? Sei p_0^s die Zahl der vakanten Zellen, p_1^s die Zahl derer, die ein Quant enthalten, p_2^s die Zahl der Zellen, die zwei Quanten enthalten usf. Die Zahl der möglichen Verteilungen ist dann

$$\frac{A^s!}{p_0^s!\,p_1^s!\ldots}, \quad \text{wobei} \quad A^s = \frac{8\pi\nu^2}{c^3}\,d\nu^s,$$

und wobei

$$N^s = 0\cdot p_0^s + 1\cdot p_1^s + 2p_2^s \ldots$$

die Zahl der zu $d\nu^s$ gehörigen Quanten ist.

Die Wahrscheinlichkeit des durch sämtliche p_r^s definierten Zustandes ist offenbar

$$\prod_s \frac{A^s!}{p_0^s!\,p_1^s!\ldots}.$$

Mit Rücksicht darauf, daß wir die p_r^s als große Zahlen betrachten können, haben wir

$$\lg W = \sum_s A^s \lg A^s - \sum_s \sum_r p_r^s \lg p_r^s,$$

wobei

$$A^s = \sum_r p_r^s.$$

Dieser Ausdruck soll ein Maximum sein unter der Nebenbedingung

$$E = \sum_s N^s h\nu^s; \quad N^s = \sum_r r\,p_r^s.$$

Die Durchführung der Variation liefert die Bedingungen

$$\sum_s \sum_r \delta p_r^s (1 + \lg p_r^s) = 0, \qquad \sum_s \delta N^s h\nu^s = 0$$

$$\sum_r \delta p_r^s = 0 \qquad \delta N^s = \sum_r r\,\delta p_r^s.$$

Hieraus folgt

$$\sum_s \sum_r \delta p_r^s (1 + \lg p_r^s + \lambda^s) + \frac{1}{\beta} \sum_s h\nu^s \sum_r r\,\delta p_r^s = 0.$$

Daraus folgt zunächst

$$p^s_r = B^s e^{-\frac{r h \nu^s}{\beta}}.$$

Da aber

$$A^s = \sum_r B^s e^{-\frac{r h \nu^s}{\beta}} = B^s \left(1 - e^{-\frac{h \nu^s}{\beta}}\right)^{-1},$$

so ist

$$B_s = A^s \left(1 - e^{-\frac{h \nu^s}{\beta}}\right).$$

Ferner hat man

$$N^s = \sum_r r p^s_r = \sum_r r A^s \left(1 - e^{-\frac{h \nu^s}{\beta}}\right) e^{-\frac{r h \nu^s}{\beta}}$$

$$= \frac{A^s e^{-\frac{h \nu^s}{\beta}}}{1 - e^{-\frac{h \nu^s}{\beta}}}.$$

Mit Rücksicht auf den oben gefundenen Wert von A^s ist also

$$E = \sum_s \frac{8 \pi h \nu^{s^3} d \nu^s}{c^3} V \frac{e^{-\frac{h \nu^s}{\beta}}}{1 - e^{-\frac{h \nu^s}{\beta}}}.$$

Mit Benutzung der bisherigen Resultate findet man ferner

$$S = k \left[\frac{E}{\beta} - \sum_s A^s \lg\left(1 - e^{-\frac{h \nu^s}{\beta}}\right)\right],$$

woraus mit Rücksicht darauf, daß $\frac{\partial S}{\partial E} = \frac{1}{T}$, folgt, daß $\beta = kT$. Setzt man dies in obige Gleichung für E ein, so erhält man

$$E = \sum_s \frac{8 \pi h \nu^{s^3}}{c^3} V \frac{1}{e^{\frac{h \nu^s}{kT}} - 1} d\nu^s,$$

welche Gleichung Plancks Formel äquivalent ist.

(Übersetzt von A. Einstein.)

Anmerkung des Übersetzers. Boses Ableitung der Planckschen Formel bedeutet nach meiner Meinung einen wichtigen Fortschritt. Die hier benutzte Methode liefert auch die Quantentheorie des idealen Gases, wie ich an anderer Stelle ausführen will.

Contributors

KIRAN BEDI, Inspector General of Police, Joint Commissioner of Police (Training), Police Training College, New Delhi. Dr. Bedi was the first woman to join the Indian Police Service. Earlier she was Inspector General of Prisons of the Tihar Jail in Delhi and later Special Secretary to the Lt. Governor of Delhi. Dr. Bedi was awarded the Ramon Magsaysay Award for Government Service in 1994.

KARSTEN BRENNER, Director General for European and International Cooperation in the German Ministry of Education and Research is a specialist in the economics of research. He served in the office of Chancellor Willy Brandt from 1969 to 1972. After six years in the Federal Ministry of Health he joined the Ministry of Research and Technology in 1978, now renamed Ministry of Education and Research. Mr. Brenner is a member of the Indo-German Consultative Group.

PETER ENGELMANN, former Coordinator of the German Ministry of Education and Research for bilateral cooperation in science and technology with India. He has specialized in fast reactor physics and worked at the Nuclear Research Centre Karlsruhe from 1960 to 1975. He was Member of the Managing Board of the Nuclear Research Centre Jülich and returned to the Research Centre Karlsruhe in 1986. He has authored a large number of scientific publications.

MURLI MANOHAR JOSHI, Minister for Human Resource Development and Science and Technology and Ocean Development since 1998. He was Professor of Physics and Head of the Department of Physics at Allahabad University till 1994. Dr. Joshi was elected to the Lok Sabha in 1977, to the Rajya

Sabha in 1992 and re-elected to the Lok Sabha in 1996 and 1998. He was President of the Bharatiya Janata Party from 1991–3. His publications include *Vikalp* (The Alternative) and *Pragya Pravah* (The Continuing Wisdom), both in Hindi, and numerous research papers.

HEINRICH F. VON LERSNER is Hon. Professor at the Technical University Cottbus. For twenty years Dr. von Lersner was the President of the Federal Environmental Agency, an institution under the responsibility of the German Ministry of Environment, which develops and implements policies for the protection of the environment on the basis of scientific criteria within the framework of a democratic polity. Among his numerous publications are: *Ökologische Wende* (Ecological Turning Point), Berlin 1991 and as co-author: *Handwörterbuch des Umweltrechts* (Manual of Environmental Law), 2 vols. 2nd ed., Berlin, 1994.

REIMAR LÜST is Professor at the University of Hamburg. Earlier he was director of the Institute of Extraterrestrial Physics at the Max-Planck-Institute for Physics and Astrophysics at Garching. He served as President of the Max-Planck-Gesellschaft from 1972 to 1984. In the following years, 1984 to 1990, Prof. Lüst headed the European Space Agency (ESA) in Paris as Director General and until 1999 the Alexander-von-Humboldt Foundation as President. Following the succesful Giotto Space Mission he was named Science Personality of the Year 1986. He is a member of numerous international science academies. His publications include: *Die Wissenschaft und die Zukunft des Menschen* (Science and the Future of Mankind), 1965, *Weltraumforschung und Kosmologie* (Space Research and Cosmology), 1970, and many learned articles.

R.A. MASHELKAR is Director General, Council of Scientific and Industrial Research and Secretary, Department of Scientific and Industrial Research. He taught Chemical Engineering at the University of Salford in Great Britain from 1969 to 1976 before joining the National Chemical Laboratory at Pune, where he later served as Director from 1989 to 1995. Dr. Mashelkar has published a large number of scientific papers, among others 'CSIR 2001: Vision and Strategy', 1996. He is a Fellow of the Royal Society, London since 1998.

GISBERT F. ZU PUTLITZ is Professor of Physics at Heidelberg University. He was Rector (Vice-Chancellor) of Heidelberg University from 1983 to 1987. He received the Leo Baeck Prize in 1989. Prof. zu Putlitz was a Visiting Professor at Yale University and at the Los Alamos Science Laboratory in USA. He has published *Atomic Physics* (together with E.W. Weber and A. Winnacker), 1975, and with B. Povh *Particles and Nuclei*, 1989, and numerous articles.

U.R RAO is Dr. Vikram Sarabhai Distinguished Professor at the Department of Space, Bengalore and a member of the Indian Space Commission. Until 1994 he was Chairman of the Indian Space Research Organization ISRO and Secretary of the Indian Department of Space. Prof. U.R. Rao is Chairman of the UN Committee on the Peaceful Uses of Outer Space (UN-COPUOS), and was President of the Third UN Conference on the Exploration and Peaceful Uses of Outer Space (UNISPACE III) in Vienna in 1999. He has published numerous studies and in 1996 his book *Space Technology for Sustainable Development*.

GERHARD REMPE is Professor in Experimental Physics at the Max-Planck-Institut für Quantenoptik at Garching since 1999. Earlier he was Professor of Physics at the Univerity of Constance. He is a member of a research team which carried out succesful experiments to verify Bose-Einstein condensation.

UWE THOMAS is State Secretary in the German Ministry of Education and Research. He is a computer engineer by training and his professional contributions are manifold in management, administration, and in politics. He served as project manager in the Heidelberg Systems Research Study Group. Later he worked in the Planning Group in the Chancellor's Office under Willy Brandt from 1971 to 1972 before joining the Federal Ministry of Research and Technology, where he was Director in charge of information technology from 1973 to 1988. Dr. Thomas was appointed State Secretary and later Minister of Economics, Technology, and Transport in the Government of the Land Schleswig-Holstein 1988 to 1993. After another spell in management he joined the Federal Ministry of Education and Research again in 1998, this time as State Secretary. His publications include *Dichtung und Wahrheit in der Verkehrstechnik* (Facts and Fiction in Transport Technology), 1996, and with P. Glotz: *Das dritte Wirtschaftswunder*. (The Third Economic Miracle), 1994.

KLAUS DIETER VÖHRINGER is Member of the Board of Management of Daimler-Benz AG (now DaimlerChrysler) in charge of Research and Technology. He joined the Daimler-Benz AG in 1967 as an engineer. Later he held important management positions in several Daimler-Benz factories before he joined the department of product planning at the firm's headquarters in Stuttgart. Since 1997 Prof. Vöhringer is Member of the Board of Management. Since 1999 he is Hon. Professor at the Technical University Darmstadt. He is also Chairman of the Committee for Research and Technology in the Bundesverband der Deutschen Industries, BDI (German Association of Industries).

STEPHAN F. VON WELCK is First Counsellor for Science, Research and Technology at the Embassy of Germany in New Delhi. Earlier he was a Senior Fellow at the German Society for Foreign Affairs (DGAP) in Bonn and Secretary General of the German-French Centre for the Conservation of Historical Monuments in Paris. His publications include *Weltraum und Internationale Politik* (Outer Space and International Politics), Munich, 1987 and *Law of Outer Space*, Baden-Baden, 1987.

Index